The Tyranny of Numbers

The Tyranny of Numbers

Mismeasurement and Misrule

Nicholas Eberstadt

The AEI Press

Publisher for the American Enterprise Institute
WASHINGTON, D.C.

1995

Available in the United States from the AEI Press, c/o Publisher Resources Inc., 1224 Heil Quaker Blvd., P.O. Box 7001, La Vergne, TN 37086-7001. Distributed outside the U.S. by arrangement with Eurospan, 3 Henrietta Street, London WC2E 8LU England.

Library of Congress Cataloging-in-Publication Data

Eberstadt, Nick, 1955-
The tyranny of numbers: mismeasurement and misrule / Nicholas Eberstadt.
p. cm. — (AEI studies: 528)
Includes bibliographical references.
ISBN 0-8447-3763-1. 0-8447-3764-x (pbk.)
1. Social sciences—Statistical methods. 2. Statistics. 3. Policy sciences. I. Title. II. Series.
HA29.E25 1995
300'.72—dc20

92-4238
CIP

1 3 5 7 9 10 8 6 4 2
Copyright credits appear on page 305.

Printed in the United States of America

To Mary
For all the innumerable reasons

Contents

FOREWORD *Daniel Patrick Moynihan* xvii

ACKNOWLEDGMENTS xxi

INTRODUCTION 1
A Global Transformation 3
The Expanding State 9
Misrule by the Numbers 15
Moral Reasoning 26

1 ECONOMIC AND MATERIAL POVERTY IN MODERN AMERICA 27
Health and Poverty 28
Infant Mortality 29
Health and Welfare 34
Nutritional Well-Being 35
Physical Data on Nutrition 38
Prosperous Paupers 41

2 THE U.S. INFANT MORTALITY PROBLEM IN AN INTERNATIONAL PERSPECTIVE 43
Historical Perspectives 44
Mortality Structures 48
Reliability of Data 49
Poverty 51
Medical Care and Perinatal Survival 54
Biological and Behavioral Factors of Low Birthweight 56
Illegitimacy 57
Parental Practices and Infant Health 60
U.S. Illegitimacy in International Perspective 62
Data at the Local Level 65

Policy Interventions and Their Limits: Income Support, Medical Care 69
Concluding Observations 73

3 HEALTH, NUTRITION, AND LITERACY UNDER COMMUNISM 74
Health 76
Nutrition 87
Literacy 89
An Irresolvable Contradiction 91

4 THE DECLINE OF PUBLIC HEALTH IN EASTERN EUROPE, 1965–1985 92
Dimensions of Mortality Change 93
Changing Cause-of-Death Patterns 101
Potential Factors in Recent Eastern European Health Problems 107
Health Care 114
Conclusions 118

5 DEMOGRAPHIC FACTORS IN SOVIET POWER 120
Fertility 121
Mortality 126
Labor Force 130
Conclusions 135

6 THE CIA'S ASSESSMENT OF THE SOVIET ECONOMY 136
Use of Soviet Data 138
Valuation of Soviet Output 141
Robustness of Estimates 143
Benefits of Review 146

7 POVERTY IN SOUTH AFRICA 150
Population 152
Health 157
Nutrition 163
Literacy 165
Conclusion 166

8 Another Look at the World Food Problem 170
Data Limitations 170
Dimensions of the Malnutrition Problem 172
Variability and Instability in the World Food System 180
Reliability of the World Food System 185
Exceptions to Global Trends 189
Conclusions 195

9 Investment without Growth, Industrialization without Prosperity 197
Industry, Investment, and Personal Consumption 199
International Capital Flows 205
Conclusions 210

10 The Debt Bomb and the World's Children 212
Resources from Poor to Rich? 214
Setbacks in Health and Education? 220
Adjustment Policies 227
Third World Priorities 236

11 World Population Trends and National Security 239
Growth in Low-Income Countries 243
Aspects of Population Change 250
Population and the Impending Challenge 264

Notes 271

Index 291

About the Author 303

Credits 305

Tables
1–1 Expectation of Life at Birth, 1940–1983 30
1–2 Infant Mortality Rates, 1940–1983 31
1–3 Infant Mortality Rates per 1,000 Live Births and Estimated Poverty Rates in Percentages for Related Children under Eighteen, 1960–1983 32

1–4 Infant Mortality Rates for Legitimate and Illegitimate Births, 1960 33
1–5 Health Characteristics of Infants in Welfare Families in the District of Columbia, 1983 35
1–6 Expenditures on Food and Beverages for Families of City Wage and Clerical Workers, 1888–1961 36
1–7 Households Receiving Less than 100 Percent of Recommended Dietary Allowance, Spring 1977 38
1–8 Estimated Daily Caloric Intake of U.S. Population and of Poverty-Level Population, 1976–1980 39
1–9 Persons without Telephones, November 1983 41
2–1 Reported Infant Mortality and Estimated per Capita Output in the United States and Fifteen Other OECD Countries, 1925–1929 and 1986/87 45
2–2 Reported Rates for U.S. Infant Mortality Compared with Reported Rates for Other Current OECD Countries, 1935–1987 47
2–3 Selected Countries Reporting Lower Mortality Rates and Lower Total Life Expectancy at Birth than the United States, circa 1985 49
2–4 Infant Deaths during the First Day of Life in the United States and Selected Other Countries, 1985 51
2–5 Infant Mortality Rates and Estimated Rates of Child Poverty in Selected Industrialized Countries, 1979–1981 53
2–6 Actual and Adjusted Perinatal Mortality Rates for Japan, Norway, and the United States, 1980–1982 55
2–7 Estimated Proportion of Low-Birthweight Babies, by Material Characteristics, 1982 56
2–8 Proportion of Low-Birthweight Babies by Marital Status of Mother, 1987 58
2–9 Infant Mortality Rates by Race, Education, and Marital Status for Women Aged Twenty or Older in Eight Pilot States, 1982 U.S. Birth Cohort 59
2–10 Infant Mortality Rates by Race of Mother and Number of Prenatal Visits in the United States, 1983 Birth Cohort 61
2–11 Percentage of Single-Parent Families among Family Households with Children in Selected Countries, 1982–1988 63
2–12 Percentage of Children in Single-Parent Families in Selected Industrialized Countries, 1960–1986 64

2–13 Estimated Average per Capita Personal Income in States and District with Highest and Lowest Reported Rates of U.S. Infant Mortality, 1984–1986 66
2–14 Illegitimacy Ratios and Proportion of Children Receiving AFDC Payments in States and District with Highest and Lowest Reported Rates of U.S. Infant Mortality, circa 1985 67
2–15 Economic Characteristics of States and District with Lowest and Highest Reported Rates of U.S. Infant Mortality, 1979–1981 68
2–16 Characteristics of Children, Families, and Persons in States and District with Highest and Lowest Reported U.S. Infant Mortality Rates among Blacks, circa 1980 70
2–17 Average Annual Number of Physician Contacts for U.S. Children under Eighteen Years of Age by Selected Characteristics, 1985–1987 71
2–18 Patterns of Consumer Expenditure by Reported Household Income and Age of Head of Household, 1988 72
3–1 Reported Increases in Mortality by Age Group in the Soviet Union, 1961–1976 77
3–2 Declines in Life Expectancy at Birth in Eastern Europe, 1964–1984 78
3–3 Mortality Change for Adults in the USSR, Eastern Europe, and Selected Western Europe NATO Countries, 1960s–1980s 79
3–4 Estimated Indexes of Economic and Demographic Change in the People's Republic of China during the "Period of Readjustment," 1978–1982 84
3–5 Reported Incidence of Selected Diseases in Cuba, 1970–1985 86
3–6 Estimated Fraction of Total Caloric Availability Derived from Grain in the People's Republic of China and Selected Other Areas, 1952–1981 88
4–1 Declines in Life Expectancy at Birth in Eastern Europe and the USSR, 1964–1985 93
4–2 Recorded Infant Mortality Rates in Eastern and Western European Countries, 1960–1985 95
4–3 Life Expectancies at Age One in Eastern and Western European Countries, 1960–1985 97

4–4 Life Expectancies at Age Thirty in Eastern Europe, USSR, and Western European Countries, Mid-1960s and Mid-1980s 98
4–5 Changes in Mortality Rates for Adults in Eastern Europe, 1965–1985 100
4–6 Changes in Mortality Rates for Adults in the German Democratic Republic and the Federal Republic of Germany, 1965–1985 101
4–7 Age-standardized Death Rates for Selected Causes in Eastern and Western Europe, circa 1985 (European Model) 103
4–8 Age-standardized Death Rates for Selected Causes in the German Democratic Republic and the Federal Republic of Germany, circa 1985 (European Model) 104
4–9 Age-standardized Death Rates for Selected Causes in Yugoslavia and Selected Southern European Countries (Greece, Portugal, Spain), circa 1985 (European Model) 105
4–10 Age-standardized Death Rates for Selected Causes in Eastern and Western Europe, 1955–79 to 1975–79 (European Model) 106
4–11 Age-standardized Death Rates for Selected Causes in the German Democratic Republic and the Federal Republic of Germany, 1975–1979 108
4–12 Age-standardized Death Rates for Selected Causes in Yugoslavia and Selected Southern European Countries (Greece, Portugal, Spain), 1965–69 to 1975–79 (European Model) 109
4–13 Estimated Annual Cigarette Consumption per Person Aged Fifteen or Older in Eastern and Western European Countries, 1965–1987 111
4–14 Estimated per Capita Annual Consumption of Distilled Spirits in Eastern Europe, USSR, and Western Europe, 1960–1980 112
4–15 European Countries with Distilled Spirits as More than One-Third of Total Consumption of Alcohol, 1980 113
4–16 Medical Personnel in Eastern and Western Europe, 1960–1985 116
4–17 Resource Allocation to Health Sector by Various Measures in Eastern and Western Europe, 1965–1985 117

4–18 Mortality Changes and Official Measures of Economic Change in Eastern and Western Europe, 1955–1985 118
5–1 Population Growth, 1950–1985, and Doubling Times at 1979–1985 Growth Rates in the USSR and Its Republics 121
5–2 Fertility Rates in the USSR and Its Republics, 1958–1983 123
5–3 Crude Birth Rates in the USSR and Selected Republics, 1950–1980 124
5–4 Projected Population under Five Years of Age in the USSR and Selected Republics and Regions, 1985–2000 125
5–5 Changes in Death Rates at Specific Ages in the USSR, 1964–1985 126
5–6 Life Expectancies at Birth in the USSR, 1958–1986 127
5–7 Ratio of Incidence of Infectious Diseases, the USSR to the United States, 1970 and 1979 128
5–8 Life Expectancies at Birth in the USSR and Its Republics, 1975 and 1983 129
5–9 Working-Age Population of the USSR and Average Annual Growth Rate, 1950–1985 131
5–10 Labor Force Participation Rates in the USSR, 1971 and 1981 132
5–11 Labor Force Participation Rates in the USSR, United States, West Germany, and Japan, 1980 132
5–12 Projected Net Changes in Population of Working Ages in the USSR and Selected Republics and Regions, 1980–2000 133
5–13 Urban Population in the USSR and Its Republics, 1950–1980 134
6–1 Estimated Meat and Milk Production in the United States and the USSR, 1960–1988 139
6–2 Estimated Changes in Aggregate Factor Productivity, per Capita Consumption, and Share of Consumption in the Soviet Economy, 1961–1987 142
6–3 Mortality Rates in the USSR and Selected Latin American Countries, 1985–1988 144
6–4 Estimated per Capita Output for Selected European Countries and the USSR, 1985 145
6–5 Estimated Real Economic Growth in the USSR and Selected Countries, Western Europe, 1981–1988 148

7–1 Population by Racial Designation in South Africa, 1985 153
7–2 Estimated Geographical Distribution of Black Population in South Africa, 1960–1985 154
7–3 Estimated per Capita Real Disposable Income, by Racial Designation and Location in South Africa, 1980 and 1983 155
7–4 Life Expectancies at Birth by Racial Designation in South Africa, 1935–1981 158
7–5 Infant Mortality Rates by Racial Designation in South Africa, 1940–1985 159
7–6 Estimated Economic Output and Structure in South African Homelands and Selected Sub-Saharan States, Early 1980s 167
8–1 Life Expectancies at Birth in Developing Regions or Countries, 1955–1980 177
8–2 Estimated Changes in Infant Mortality Rates in Developing Countries, 1955–1980 178
8–3 Estimated Total Output of Agricultural Products and Food in Sub-Saharan Africa, 1979–1981 192
9–1 Changes in Estimated Life Expectancy at Birth in Selected Regions, 1950–1985 198
9–2 Changes in Estimated Infant Mortality Rates in Selected Regions, 1950–1985 199
9–3 Estimated Industrial Production and Gross Domestic Investment for Selected Countries and Groups, 1985 200
9–4 Net Western Financial Flows to Developing Countries, 1956–1986 206
9–5 Structure of Net Western Financial Flows to Selected Classifications of Countries, 1977–1986 208
10–1 Estimated Net Financial Transfers to Less-developed Countries, 1980–1987 217
10–2 Concessionality in Long-Term Lending to Less-developed Countries by Interest Costs versus LIBOR-Dollar One-Year Rates, 1980–1988 219
10–3 Estimated Infant Mortality Rates, 1980–1985 222
10–4 Reported Rates of Infant Mortality for Non-Communist Less-developed Countries, 1980–1988 223
10–5 Estimated Net Enrollment Ratios by Age Group for Children in Developing Countries, 1980–1987 226

10–6 Estimated Central Government Expenditures plus Lending, minus Repayments, as a Percentage of GDP for Selected Regions and Countries, 1980–1988 230
10–7 Estimated Allocation of Central Government Expenditures to Health and Education for Selected Groups and Countries, 1980–1988 232
10–8 Estimated Annual Growth in Government and Private Final Consumption Expenditures for Selected Regions, 1980–1986 233
11–1 Changing Projections for Population of Selected Regions in the Year 2000 242
11–2 Estimated Total and per Capita Output and Growth for Selected Areas, 1900–1987 245
11–3 Estimated Postwar Health Progress in Low-Income Areas of Selected Regions, 1950–1985 246
11–4 Population Aged Sixty-Five and Older in Selected Industrial Economies, 1950–2025 252
11–5 Russian and Central Asian Population of the USSR, 1979–2050 258
11–6 Most-Populous Countries, 1950–2025 265
11–7 Changing Global Population Balance for Selected Groupings and Countries, 1950–2025 267
11–8 Changing Distribution of Global Births for Selected Groupings and Countries, 1950–2025 268

FIGURES

1 Total Outlays of Government as a Percentage of GDP in the G-7 Countries and Smaller OECD Countries, 1960–1992 14
2 Consumption versus Income-based Measures of Poverty in the United States, 1959–1989 18
8–1 Production Shortfalls and Price Changes in the Global Grains Economy, 1961–1991 182

Foreword

Somewhere in the works of George Orwell there is an account of a reaction to the arrival of American troops in Great Britain in the middle of World War II. It would seem the rumor went round certain "right left" circles, as Orwell called them, that the Yanks had not come to help mount an invasion of the Continent. Rather, they had come to put down an incipient workers' revolt in the war industries—some vague allusion to Petrograd in 1917. Orwell commented that in order to believe such an absurdity, it was necessary to have gone to a university. Any cabdriver, he continued, would have told you it was nonsense.

So would Nicholas Eberstadt, who in our time is becoming the counterpart of Orwell's cabdriver (albeit his subject is as often the inanities of the political right as of the left). In our more democratic setting, he is assuredly university-educated. He is currently a visiting fellow of the Harvard University Center for Population and Development Studies and a visiting scholar at the American Enterprise Institute for Public Policy Research. But he has retained and indeed developed an uncanny ability to ask the innocent, devastating question; to spot the stunningly illuminating obvious.

Eberstadt is by training a demographer and would not, I surmise, dissent from the stern assessment that demography is destiny. It is all the more important then for those who would know the future to get their numbers straight. Early in his academic training and his published works, Eberstadt kept spotting instances where people who had every reason to get things straight got them all mixed up instead. He acquired a mission of sorts: asking the obvious.

I will not ever forget a meeting with him in my Senate office early in 1990. I was then trying to think through a set of hearings on American foreign policy at the end of the cold war. It appeared to me that the collapse of the Communist regimes in Eastern Europe—the Berlin Wall—had taken official Washington quite by surprise. The turmoil and tumult in Moscow seemed equally unanticipated and incomprehensible to our official analysts. I wondered, Were we going to leave in place, unexamined, unreformed the institutional arrangement and patterns of thought that had led us to "miss" the biggest event of the century? But how to make this point? Enter Eberstadt.

He had asked if he might bring around a group of visiting Soviet economists. Most assuredly, I replied. It seemed important at the time and it was important to be welcoming as Soviet intellectuals made their first tentative probes of Western opinion, asking, in effect, what should they do now? Young Eberstadt arrived with a half-dozen Muscovites. They had a story they wanted to tell. The story was that the Soviet economy was already in ruins, and was heading for worse. It had never been anything as large as it was thought to have been, or was said to have been, by them or by us. And now it was going to get worse, worse to the point of instability.

At this time the Central Intelligence Agency was estimating the size of the Soviet economy at more than half that of the economy of the United States. Estimates of growth rates in the USSR had been scaled back. (Years earlier, as a skeptical Walt Rostow at the time remarked, we had pretty much bought into a "6 percent forever" school.) But even so, what the United States government "knew"—what the president of the United States "knew"—was that the Marxist economy of the Soviet Union had transformed that nation from an agricultural backwater into a modern industrial state, capable of sustained growth to ever-higher levels of economic achievement. Our new Soviet visitors fair to pleaded: "That is not so. That was never so."

I listened with attention and just a touch of caution. Were they thinking Marshall Plan? The "soft" loan window at the World Bank? Were the official numbers *that* wrong? At this point Eberstadt reached into his book bag and fetched up the 1989 edition of the *Statistical Abstract of the United States,* a Bureau of the Census publication, a hugely useful compendium that draws on data provided by all manner of U.S. government agencies, including the Central Intelligence Agency, which by this time was publishing some of its estimates. These included the estimated size of the economies of the Federal Republic of Germany and the German Democratic Republic, West Germany and East Germany as they were known. And there it was!

To wit, per capita output in 1985 in East Germany was shown as fully equal to that of West Germany. In fact, the volume showed per capita GDP in East Germany as ever so slightly higher! These figures, Eberstadt explained, had come from an unclassified and widely used CIA publication. Of a sudden, all was clear. Any Berlin taxi driver crossing over at Checkpoint Charlie could have told you that the economy of West Germany was manifestly superior to that of East Germany. Somehow, however, what Thorstein Veblen (in *The Instinct of Workmanship*) called "trained incapacity" in the most exclusive and hidden regions of the American government had caused us to get it all wrong. *Our* taxi driver, Nicholas Eberstadt, spotted it.

I invited him, of course, to testify before the Committee on Foreign Relations in a series of hearings that took place in the summer of that year. His superb statement can be found as chapter 6 of this volume. I especially commend to readers this one passage: "It is probably safe to say that the U.S. government's attempt to describe the Soviet economy has been the largest single projection of social science research ever undertaken."

And now we know that that project was a huge failure—and a hugely expensive one for us, or at least that can be argued. (Had we known how weak the Soviet economy was, would we have been more careful of our own resources? Without wishing to intrude my own thoughts, it is a matter of record that in 1979 I argued that the 1980s would see the breakup of the Soviet Union. This argument was incomprehensible to official Washington, not least the official intelligence agencies. It was not thought worthy even of rebuttal.)

Eberstadt's method is, of course, much more subtle than at first appears. His questions are so *simple*. It is only when the answers turn out to be anything but simple that the reader recognizes the extraordinary analytic powers of this most welcome member of the new generation of American social scientists: the more welcome for the unobtrusive but unmistakable moral dimension of his work. His work is not that of a disembodied analyst. He is a demographer. The *demos* is his setting, be it ever so exotic—or so close to home as to seem familiar. This volume begins with an essay entitled "Economic and Material Poverty in Modern America." It is full of *seemingly* familiar material, until, in the penultimate paragraph, Eberstadt raises the issue of child poverty in the United States that a quarter-century of commissions and panels, committees and boards have contrived to avoid at whatever cost. This is the scarlet fact of pauperism.

> The results of American social policy over the past generation, for all the importance of . . . physical successes, do not im-

> mediately excuse the government, or the public that sustains it, from the criticism that in eliminating physical poverty we have created in our midst a class of prosperous paupers. The term "pauperism" is lost on the modern ear, but its meaning, as given by the *Oxford English Dictionary*, is specific: "dependence upon public relief, as an established condition of fact among a people."

By our calculations in the Subcommittee on Social Security and Family Policy of the Senate Committee on Finance, nearly one-third of children (31 percent) born in the United States in 1980 will have been on Aid to Families of Dependent Children by age eighteen. This is somewhat higher than the historical ratios, which are now known, but about where we have been for a generation. These children, however much headstart they get, however much nutritional supplements they receive, these children are paupers. If you would rather not think about this, stay clear of Eberstadt. But if you have the heart for it, here is your man!

DANIEL PATRICK MOYNIHAN
U.S. Senate

Acknowledgments

These essays were written during years in which it was my great good fortune to be associated with both the Harvard Center for Population and Development Studies and the American Enterprise Institute for Public Policy Research.

Having been a member of the Harvard Center, and a beneficiary of its encouragement and support, for more than fifteen years, I am keenly aware that my debts there are too great to enumerate. Even so, I would be seriously delinquent if I failed to mention my gratitude to the center's current director, Professor Lincoln C. Chen, who has striven to make it the exciting place that it is today.

My debts to AEI predate its current president, Christopher DeMuth, but he has only added to them over the past eight years. For despite his impossible schedule and the endless demands on his attention and patience, he has always found the time to take a personal interest in my research projects and has made it his business to help with each new undertaking I have proposed.

During the years in which this book has been in progress, it has also been my good luck to have the help of two outstanding research assistants: Elizabeth Blackshire and Jonathan Tombes. Both Miss Blackshire and Mr. Tombes have distinguished themselves on their tours of duty under my command. Not only have they been exacting, meticulous, versatile, and tireless, but they have somehow managed to be ceaselessly cheerful all the while. They and I know this book would not yet be ready but for their efforts.

Only a few years ago, when the early versions of the essays in this book first began to appear, many of the arguments and findings were considered controversial. There seems to be rather less disagreement about some of these points today. Personally gratifying as this may be, it only deepens by own gratitude to the benefactors who had the confidence in my work to see it through. The Smith Richardson Foundation and the Lynde and Harry Bradley Foundation specifically sponsored two of the studies that appear in this volume. The Sarah M. Scaife Foundation has been a general and abiding supporter of my research at AEI. I owe a special thanks to the John M. Olin Foundation. At a critical moment, when my funding seemed most uncertain, the Olin Foundation moved quickly, and generously, to underwrite my research.

Chapter 8, "Another Look at the World Food Situation," was begun in collaboration with my old friend Clifford M. Lewis. It is better than it otherwise would have been for the insights and observations that he has contributed.

My greatest debt of all is to Mary. In every journey that this volume chronicles, she has been my companion. It is she who first read these chapters; she who supplied the most penetrating critiques and the most helpful suggestions; she who suggested the title. In dedicating this book to her, I hope only to acknowledge an obligation. The obligation itself is far too vast to be redeemed.

The Tyranny of Numbers

Introduction

Most of us probably regard the continuous collection of official statistics by governments nowadays as a dull but essentially harmless activity. The facts and figures that modern governments amass may well appear dull, but it is my contention they are not essentially harmless. To the contrary: in this collection of studies I attempt to demonstrate that ordinary people around the world routinely suffer injury through the agency of these selfsame dull statistics. On more than a few occasions, these injuries have been grave and irreversible and have afflicted large numbers of persons.

This peculiar and modern form of political abuse, as we shall see, is not solely the misfortune of benighted peoples in distant lands who strain under the rule of unaccountable despots. As the following chapters argue, similar if less extreme problems regularly burden the citizens of affluent and open societies with freely elected governments. Indeed, it is my further contention that the injury and damage wrought on individuals and even entire populations through the official use of official data can only be mitigated—never entirely eliminated. This is an unpublicized but nonetheless real consequence and cost of the political arrangements we have selected for ourselves. Though he may not always recognize his bondage, modern man lives under a tyranny of numbers. And if my analysis is correct, this tyranny of numbers is a foretaste of what we may expect from the future.

Once invested with political power, statistics can injure human beings. Modern governments—rationalist and problem solving as they are in orientation—imbue statistics with political power by charting their policies against them. The power accorded to these mute and inanimate numbers today is far-reaching; it can be felt around the world, regardless of a state's professed philosophy or ideology. This should not surprise, for what makes a state modern, in an important sense, is its intention to use numbers as a basis for its actions. The modern state is an edifice built on numbers.[1] Modern governments, unlike the diverse governments of earlier times, require statistical information simply to function *as* modern governments: to perform the tasks now conventionally assigned to and expected of them.

Under modern governance, however, the impetus for state action typically extends well beyond the limits of knowledge that available data are providing—or, indeed, could ever possibly provide. This is the perilous terrain where the tyranny of numbers prevails. In this dangerous realm, government decisions confer benefits on the citizenry only by chance since they cannot do so by realistic design. Within this realm, every individual becomes a potential victim of misrule—a hostage to chance whose fortunes and fate may ultimately be determined by misguided government policies or state practices.

One may object that this reading of the civil condition romanticizes the past and demands too much from the present. After all, one may observe, governments have *always* had to operate under conditions of uncertainty, and the governments of yesteryear were generally rather *less* capable of coping with uncertainties than the states that exist today. Moreover, the information revolution over the past several generations has, in fact, pushed out the frontiers of uncertainty facing the statesman and the policy maker far beyond any perimeter that might have been imagined in the past—much to the advantage of the public at large.

Let me be clear. I entertain no particular fondness for any of the illiberal political arrangements from bygone times, most of whose shortcomings are all too clear today. From a purely material standpoint, mankind is much better off today than ever before, as I hope the following chapters attest. What I wish to suggest in the following pages is that the statistics-oriented, problem-solving, meliorative state, while vastly preferable to many variants of rule previously experienced, nevertheless poses new and unfamiliar hazards to mankind.

Where unshakable traditional beliefs or passing superstitions played official roles in the past, we now witness overconfidence based on a false precision. Where innocents lost their freedom, or their lives, through the state-directed autos-da-fé of yesteryear, blameless citizens and subjects are now punished by their governments' routine misuse of the techniques of social science (an act of faith of a secular sort that as yet attracts little comment). Where antique despots surrendered to the temptations of numerology, the modern statesman proudly succumbs to the allure of "quantophrenia"[2]—an idolatry of numbers no less unreasoning, and no less poorly suited for promoting the commonweal, than its precursor.

Yet, in addition, the conceit of the problem-solving state encourages a range and a depth of constant official interventions into daily life that would not have been contemplated by the autocracies of old; not a few of these innumerable interventions result in adverse outcomes for intended beneficiaries and in unintended costs or hardships for many other persons.

Reflecting on their personal prospects in an imperfect world, most people would have no difficulty with the proposition that some forms of tyranny are harsher than others. The restrictions on individual freedom endured, and the material losses incurred, by citizens and subjects through governments' use and misuse of statistics would hardly qualify as the harshest of the tyrannies that mankind has experienced in its long and often unhappy experiment with governance. Many, possibly most, people might opt for a tyranny of numbers over the feasible alternatives if given the choice.

Be that as it may: realism, if nothing else, requires us to call things by their proper names and to describe their properties as best we can. It is to this purpose that the collection of studies in this book is dedicated. For while the following chapters individually address distinct and seemingly separate topics within the purview of contemporary public policy, they are united by a common theme: the abiding danger that domestic and international policy will be distorted, corrupted, degraded, or perverted through a misuse of statistics and quantitative measurements.

To understand the significance of the phenomenon of misrule through numbers, and to appreciate what it may portend for populations today and in the future, we must examine both the defining political events of our century and the reasons for the inexorable rise of the statistics-oriented, problem-solving state. The next few pages of the introduction offer a brief glimpse at each of these.

A Global Transformation

For the third time in our century, the whole world has been shaken by a political earthquake. Like the two worldwide upheavals that came before it, this one has suddenly and unexpectedly transformed the entire international landscape. And like those two earlier global convulsions, this one promises to affect the way we live—wherever we may live. For the third time in our century, dramatic events around the world are raising the great and terrible question of modern politics: how are human beings to be treated by the states that preside over them?

The first of the three great upheavals that was to change the way men and women around the world were governed occurred between 1914 and 1918. The Great War, as it was once known, culminated in a radical reconfiguration of international affairs. By shattering three European empires, the Great War brought to an ignominious close the post-Napoleonic concert of nations. An equilibrium between Great Powers with explicitly limited international ambitions—that is to say, an international balance of power—was no longer practicable (nor, to many eyes, even desirable). Thus the Great War ushered in an era defined by a quest for international mastery, relentlessly pursued by new, ambitious, and indeed doctrinally insatiable powers.

If all this augured ill for the disposition of governments toward vulnerable populations, other particulars made the character of rule emerging in this postwar order look even less auspicious. In the course of mobilizing for this, the first total war, the combatant regimes discovered new administrative techniques for harnessing the subjects or citizens under their sway and generating previously unimagined power with which to pursue the purposes of state. The intellectual triumph of the theories of nationalism and self-determination served notice that human beings, in the new order, could be stripped of their rights in the land of their birth if theirs was a disapproved ethnicity or "race"; even for the favored and governing nationalities, however, the rights of power did not by these theories necessarily extend civil protections or guarantees to individual members of the group. Perhaps most ominous of all, a fundamentally new kind of state emerged from the wreckage of the Russian Empire. The Soviet government embraced a truly revolutionary political theory, succinctly summarized by its founding theorist, V. I. Lenin: "We recognize nothing private."[3]

The next great political upheaval of our century was World War II. That war is given different dates in different regions of the globe, in accordance with the duration of carnage locally, but it may fairly be said to have begun in the Far East in 1937 and to have concluded, again in the Far East, in 1945. This long and brutal struggle witnessed far-reaching innovations in the technique of the war economy and saw the advent of the atomic bomb, an entirely new order of weaponry. These breakthroughs in organization and technology promised to endow government in the postwar era with vastly more destructive force than it had ever before possessed. The character of the governments into whose hands these powers would pass was cause for growing concern. This, too, was a direct outcome of the war itself. Though France and Britain emerged from the war as technical victors, they were drained and exhausted; their days as world powers were over. For them, and for the other European countries that still held foreign territorial possessions, a great decolonization commenced. In the process, the barbaric European doctrines of nationalism and self-determination, which had already caused so much suffering on the continent of its origin, were visited anew on peoples throughout Africa and Asia.[4]

Yet the story of decolonization was only one of the dramas to frame the postwar order and by no means the dominant one. The tenor of postwar politics was established by the war's virtual victors, the United States and the Soviet Union. Constituted as they were along irreconcilable and necessarily hostile political principles, these erstwhile allies were transformed through victory into global opponents. The tense contest that ensued, with its separation of countries according to their "camp," only accentuated the growing contrast between modes of contemporary governance. There were regions of the postwar world—Japan, Italy, and West Germany among them—whose populations had never before enjoyed such general civil and political liberties. A much larger portion of mankind, however, was thrust into varieties of vassalage unique to the modern era. Indeed, never before had so much of humanity lived in such peril of, or under so complete a servitude before, what were (after all) its own governments.

The revolution in world governance that is going on all around us today is comparable in its moment with the changes ushered in by World War I or World War II. Yet our current political earthquake is fundamentally different from the two that preceded it. Unlike the two world wars, each of which purchased a restructuring of international and domestic affairs with the lives of tens of millions of soldiers and civilians, the upheavals that have recast the international order over the past few years may by any historical measure be described as nearly bloodless. And where the two world wars heralded, or ratified, the establishment throughout the world of new political systems under which the rights of the individual could never be secure, the revolutions of the recent past suddenly put all the modern varieties of autocracy on the defensive. Around the world, and in seemingly unrelated manner, despotisms devised and nurtured by the twentieth century are now crumbling or in disarray under the pressure of events.

Consider the great political transformations that have been registered in the world arena in the six years between January 1989 and December 1994. In Eastern Europe, the revolutions of 1989 toppled the seemingly solid Socialist dictatorships of the Warsaw Pact alliance as if with a sweep of the hand, permitting peoples long trapped by Soviet-style "liberation" at last to rejoin the parade of history. The year 1989 also marked the beginning of the end for the Soviet state. Though the totalitarian project in the USSR was the first of its kind, and by many measures the most successful of the many that followed it, the

Soviet Union was, by the start of its eighth decade, plummeting into a final crisis, propelled by problems of its own making. Irremediable political decay and a succession of erratic but largely cosmetic measures for redressing this decay found their synthesis in the amateurish Moscow coup of August 1991. The coup itself signaled the end for the Communist party; its failure signaled the end for the Communist state. By early September 1991, the Soviet structure was clearly doomed; it disappeared in its entirety before the year's end. Thus the single most fearsome apparatus to emerge from the twentieth century exited the political stage with a whimper, not a bang.

If the collapse of Soviet-bloc communism were the only event of note on the world stage between 1989 and 1994, there would be reason enough to rank the significance of those years on a par with World War I or World War II. But the boundaries of the former Warsaw Pact alliance do not delimit the territories in which established enemies of liberty find themselves in retreat. In diverse regions of the globe, twentieth-century despotisms of varied genealogy are being accosted from an unaccustomed direction. As our century draws to a close, these organized systems of unfreedom are being bent, and broken, by the forces of political liberalism.

South Africa has exemplified this change in the international political climate. The radical elite that was voted into power in that country's 1948 Whites-only election set itself the task of developing and perfecting a legal and administrative structure to confer preference and benefit upon but a single group in this multiethnic land. This new political order came to be known internationally by its designation in the Afrikaans tongue: apartheid. Apartheid was a modern idea, not an antique notion. The apartheid system, after all, was in its essence a faithful formalization of ideas central to the doctrine of nationalism. (This fact, incidentally, was not lost on South Africans themselves; not for nothing did the dominant and ascendant political force during the apartheid era name itself the Nationalist party.)

Yet even in the eyes of the group expressly intended to profit by it, this system of ethnically grounded restriction and repression could not maintain its legitimacy. The crumbling of the apartheid state began quietly; there followed increasingly telling indications that its policy lacked moral authority among its own ostensible constituents. A milestone came in 1986, when South Africa's Dutch Reformed Church—the faith to which the overwhelming majority of the country's Afrikaaners confessed—reversed its earlier teachings and declared that there was no scriptural basis for the policy of enforced racial separation.

The final death knell for apartheid tolled in March 1992. In an extraordinary political referendum, South Africa's Whites-only electorate voted to end White rule by a margin of more than 2 to 1 and to instruct the sitting Nationalist government to negotiate a new constitution under which civil and political rights would be guaranteed to all. In May 1994, South Africans of all ethnicities participated in the country's first one-man, one-vote elections; immediately thereafter, a peaceful transition to a multiethnic, multiparty administration was effected. Thus, scarcely a matter of months after the formal demise of the Soviet Union, another state of injustice, organized along disparate lines, was decisively and dramatically overturned.

South Africa's apartheid regime, it is true, was an atypical exemplar of the arrangements that bound non-Communist populations into vassalage during the cold war era (if only because a limited portion of the country's populace *did* actually enjoy genuine civil and political liberties). But the more familiar variants of third world autocracy—the police state and the one-party state—have recently been gripped by a crisis of legitimacy as well. Between 1989 and 1994 some of the hardiest of these states were lost—not to revolution but to peaceful evolution.

On the island of Taiwan, the Kuomintang (Nationalist party, or KMT), which had arrogated for itself a monopoly of power under "emergency" martial law legislation in place since its forced departure from the Chinese mainland, embarked on a voluntary and deliberate self-transformation. Over a period of years, this monolithic party purposely sacrificed its extralegal privilege and strove to create an open, competitive, and genuinely constitutional political framework for government. At the same time, on the other side of the world, openness and accountability were being championed by a most unlikely agent: the long-ruling Party of the Institutionalized Revolution (PRI) in Mexico. Deliberately and methodically, the PRI commenced a relaxation of its merciless grip on the nation's politics; in local and regional elections, PRI candidates no longer automatically achieved victory against other parties. No less significant, the PRI moved decisively toward restoring to Mexico's citizenry many of the economic liberties it had previously deprived them of.

Despite their reforms, neither Taiwan nor Mexico would have qualified by the end of 1994 as *Rechtsstaaten:* states restrained by, and governed under, stable rule of law. Yet, over the course of only a few years, they had moved momentously in that direction. And they were by no means the only third world autocracies to do so. Paradoxically, the collapse of the Soviet state seemed to expose *non-Communist* dictatorships in the third world to systemic risk. For with the end of the cold war, Western powers were relieved of a major reason for supporting internationally the authority of governments whose domestic practices dismayed or even disgusted them.

One should not exaggerate or idealize the dimensions, to date, of this global upheaval. Tarnished and battered by recent events as the world's remaining despotisms may be, the final triumph of liberal rule is hardly at hand. The sway of modern tyrannies remains vast. In China, the world's largest population is still beneath the hand of a ruthless Communist dictatorship; for all the economic experimentation in that land since the death of Mao Zedong, even permission to seek self-enrichment is as yet nothing more than a provisional gift from the state to the people—a gift, moreover, that could be taken back at any time by ukase. In the non-Communist world, we see governments that are conducting campaigns of mass murder both out of principle and for the utter lack of it: look only to the former Yugoslavia's monstrous "ethnic cleansing" for an instance of the former and to the terrible famine thrust on Somalia by that nation's contending warlords for an example of the latter. The slaughter in Rwanda in 1994 provides another reminder of the terrible purposes that are still being pursued by those who presume to govern—if such a reminder is needed.

Moreover, even if such barbarous cliques could all somehow be removed from power, the unsettling fact is that the liberal ideal is viewed with suspicion, if not outright hostility, by ordinary people in many regions of the contemporary globe. The current popularity of extremist movements in the Islamic expanse between Casablanca and Kabul seems to attest to nothing less.

But if the global triumph of the liberal ideal is not yet at hand, the prospect of its ultimate triumph is no longer a fantastical notion. With the end of the cold war, international relations are suddenly dominated by states embodying and championing varieties of political liberalism. The possibility that the world's *domestic* policies might someday bear this same stamp—that limited and constitutional governance, rule of law, and respect for the rights of the individual may come to typify the obligation of the standing state to its citizenry—does not seem quite so remote and quixotic as it did only a few years ago. In the wake of the great events late in our century, hardheaded realists should now join utopian dreamers in contemplating the nature of a world remade in accordance with the precepts of political liberalism, for there is more than a small chance that we or our descendants will be living in such a place. In a memorable phrase, it is a world that will have reached the "end of history."[5] So indeed, in a Hegelian sense, it will. But people will still live and breathe, and governments will still operate. It is perhaps not too soon to think about some of the problems that may await mankind after it arrives at history's end.

From the standpoint of the disfavored and the weak, a world characterized by liberal polities would surely look better than anything we have ever known before. At the same time, it would be unwise to minimize the dangers that would still lurk in such a world. For even in what, from our current vantage, might look like the best of all possible futures, mankind's long and dogged struggle for liberty will not be over. Not even the end of history promises to end the abuse and injury of human beings by reigning governments. To the contrary, it promises merely to alter the nature and to moderate the severity of these abuses and injuries: to substitute a more pleasant tyranny for the more fearsome forms we see around us today. If political liberalism prevails, our descendants will still have cause to fear the consequence of policies deliberately pursued by their own government and by others like it. For they will be familiar with a special kind of tyranny: one that we do not seem to understand today, even though it is amply evident. This is the tyranny of numbers: misrule directly born of the misuse of statistical information by problem-solving states.

The Expanding State

Ever since the Enlightenment, exponents of liberal politics have assigned the state a responsibility not only for protecting its citizens but also for improving their lot. Over the intervening generations, the realm, scope, and emphasis of those improvements envisioned by advocates of liberal governance have gradually, but unmistakably, changed.

Whereas a Hume—or a Montesquieu or a Kant—saw enlightened government as an indispensable instrument for improving the moral environment to which citizens were exposed, subsequent expositions of the liberal rationale focused increasingly on improvements in the public's *material* circumstances. In their contemplations on material advance, moreover, the intellectual successors to Bernard Mandeville and Adam Smith now take for granted the need for direct, broad, and continuous interventions by government into the economic affairs of its citizenry. Over the past two centuries, free peoples around the world have built themselves an apparatus of government that has become quite similar from one country to the next—yet would seem nearly unrecognizable to the apostles of an earlier, revolutionary liberalism. Virtually without exception, free peoples in open societies in all corners of the globe now affirm their preference for living under meliorative states: governments charged to, and empowered to, redress, or at the least address, social and economic problems confronting the populace.

Not so many generations ago, the idea of a progressive and problem-solving state would have sounded amazing and implausible to all but a tiny fringe. That such states are now commonplace attests to how much in our own world was unforeseen only a short while ago.

This transformation, like so many other transformations before it, owes much to an intellectual revolution. In just a few lifetimes, the philosophy of rationalism burst out of Europe and swept the world, conquering nearly all that did battle with it—and claiming their children in the process. Today, whether we understand so or not, we are all rationalists—even those who oppose or despise the liberal ideals. Whether we like it or not, our politics are, quite inescapably, the politics of rationalism.[6] The purposeful extension of the scope of government, along with the deliberate reconstitution of government as a problem-solving (and problem-seeking) apparatus, is entirely consistent with the rationalist disposition—some would say ordained by it. But the rise of the problem-solving state has also been propelled by considerations less cerebral and more quotidian.

For the explosive—and routinized—growth of knowledge since the early nineteenth century has radically altered the boundaries of the possible. As the frontiers of technical and administrative feasibility were extended, the obligations and expectations devolving on limited and constitutional orders were correspondingly redefined. The tasks that free peoples assigned to their public servants steadily multiplied; the magnitude and complexity of these assignments waxed as well. Though political forces and ideological passions most assuredly played a role in this worldwide expansion of limited government's purview, the expansion itself could hardly have taken place if it had not been widely judged a thoroughly practical affair.

As the twentieth century progressed, free peoples entrusted ever more assignments to the states that served them because, in the final analysis, they were reasonably confident about the results that they could expect from their public administrators. Rationalistic, systematic public policy promised the citizen protections and benefits that individuals could not secure by their own isolated actions. Moreover, meliorative government demonstrated, in one field of endeavor after another, that the hand of the state could replace the "invisible hand" tolerably well—and could actually improve upon its work on more than isolated occasions. As confidence in the meliorative properties of the problem-solving state has deepened, ever more ambitious problems have been delegated to it. It has taken the liberal democracies barely three generations, for example, to move from cautious experimentation with programs of insurance for the unexpectedly stricken, to broad measures for cushioning national populations against the swings of the business cycle, and on to regular and far-reaching policies intended to control or to reshape the business cycle itself.

The corollary to this steady expansion of assignments has been a steady expansion of the size and might of the state, even under the aegis of liberal democracy. One measure of this expansion may be divined from a simple ratio: the government's expenditures in relation to total national output. Such ratios do not fully describe the reach of the state, for they cannot speak to its legal prerogatives in civil life and need not fully reflect activities of self-financed organizations owned by the state. Additionally, these simple ratios are more difficult to compose than one might at first assume. But they are regularly assembled, and they reveal an astonishing fact: in the decade just completed, Western Europe's governments, though at peace, absorbed roughly the same fraction of national product as they did during World War II—years of total mobilization for national survival.

Consider: in 1940—the year in which its *Wehrmacht* conquered and occupied most of continental Europe—the German state's expenditures amounted to roughly 48 percent of its gross national product (GNP).[7] Forty years later, defense expenditures amounted to less than 3 percent of national output in West Germany, but general expenditures of government were estimated at slightly over 49 percent of the gross domestic product (GDP).[8] (In Austria, which had been impressed into the same *Reich* for the course of the war, defense spending amounted to barely 1 percent of GDP in the 1980s, yet general government outlays consistently exceeded 50 percent of national output throughout the decade.)[9] Indeed, available figures suggest that quite a few peaceable and open Western societies—Belgium and the Netherlands among them—entrusted a greater share of national output to their governments in the 1980s than Hitler's war state had commanded in 1941—the year of Operation Barbarossa.[10]

No less striking is the comparison between contemporary industrial democracies and their erstwhile antagonists in what was Communist Europe. Since the end of the cold war, the countries of the former Warsaw Treaty Organization have been releasing information suppressed under their *anciens régimes*. Newly available financial and commercial figures permit a clearer description of the workings of their command economies (though the effort of translating results from a central planning system into a Western market framework raises profound and even arguably irresolvable issues).

Initial recomputations by the International Monetary Fund provide estimates of their ratios of state spending to national output. According to these calculations, the differences across the iron curtain with respect to these ratios were not stark and dramatic. To the contrary: by these numbers, it would appear that many of the affluent welfare states of the West were actually allocating a greater share of their national resources through mechanisms of government than were their militarized, centralized "Socialist" counterparts in the Soviet bloc. Between

1981 and 1985, for example, general government expenditures in Hungary are now estimated to have equaled about 62 percent of GDP.[11] That is a high ratio—but over those same years, the corresponding figure in Sweden is estimated at over 65 percent. By this same reckoning, in 1988—on the eve of its downfall—Poland's Communist state was actually spending a lower share of national output (48 percent) than the governments of such places as France (49 percent) or Ireland (52 percent). And in 1989—the year the Ceauçescu dynasty was toppled—the Romanian dictatorship apparently claimed the same share of domestic product through state spending (44 percent) as Canada's constitutional, federal democracy. Though the governments in the affluent West, unlike the Communist East, were operated under the rule of law and remained accountable to their voting publics, they have indisputably become enormous and imposing edifices.

Many observers of the modern state have noted a positive correlation between its share in a country's economy and the country's income level. Such observations are generally accurate, as far as they go. In modern times, the growth of the state sector has also drawn on factors entirely independent of levels of prosperity. A comparison between contemporary India and late-nineteenth-century France should make the point. While such calculations are necessarily inexact, one recent estimate suggests that per capita output in India as of 1989 was about half of what it had been in the France of 1890.[12] Yet despite this disparity, the share of general government expenditures within the Indian economy in the late 1980s was more than *twice* as great as it had been in France a century before.[13] Late-nineteenth-century France, as it happens, is an adventitious bench mark for such comparisons. For the France of the early Third Republic, with its etatist tradition, seems to have had a larger public sector than any other country of its day.[14] Yet today scarcely any countries support public sectors so *small*, no matter how impoverished their populations.

Should we expect growth of the state sector in free societies in the years to come? Some commentators insist that we should not, adducing a variety of economic and political arguments for their case. Perhaps their predictions will be vindicated by history. But it is by no means self-evident that we have already witnessed the high-water mark for public sectors in free societies.

The American experience in the 1980s is instructive. With Ronald Reagan's overwhelming victory in the 1980 election, the presidency passed to an avowed opponent of big government. In his inaugural address, Reagan declared: "In this present crisis, government is not the solution to our problem; government is the problem. . . . It is my intention to curb the size and influence of the Federal establishment."

And, indeed, the Reagan team was widely perceived as intent on enforcing a radical reduction of the size and scope of the U.S. public sector. The Reagan administration oversaw substantial cutbacks in many social and domestic programs. These cutbacks were the focus of furious political battles and equally impassioned public commentary. Yet for all this *Sturm und Drang,* both proponents and critics of the Reagan legacy somehow manage to overlook a simple fact: the U.S. government did not shrink, but rather grew in size during President Reagan's tenure in office. In 1988—eight years into the "Reagan revolution"—both federal and total government outlays, as a share of the U.S. gross domestic product, were higher than they had been at the time of Reagan's first inaugural address.

Relentless expansion of the state sector, of course, is not the leitmotif of public finance in every contemporary Western democracy. There are several exceptions.

In the United Kingdom, a succession of resolute Conservative cabinets deliberately pressed against the share of public spending in the national economy; that ratio was, at the end of Margaret Thatcher's tenure in 1990, actually lower than it had been in the days immediately before the British general election of 1979. In Sweden, the mounting financial crisis of its welfare state compelled a temporary retreat from the arrangements of the early 1980s, when the Swedish government was spending two of every three kronor its citizens produced. In a few other member-countries of the Organization for Economic Cooperation and Development (OECD), moreover, the ratio of government outlays to national output appeared to be slightly lower in 1990 than it had been in 1980.

On the basis of these and other indications, some analysts theorize about a coming, and more general, repudiation of statism (and the political parties partial to it) by the electorates of the industrial democracies. Only time will tell whether they are correct. Such a verdict, however, would seem at least premature today.

A closer examination of public spending patterns points to a continuing growth of the state in Western societies, although the tempo of its expansion is conditioned by the vagaries of the business cycle. A comparison of 1990–1992 with 1980–1982 underscores this growth. Like 1980–1982, the years 1990–1992 saw slowdown and recession in the world economy; yet in the latter period, and in the face of what appears to have been a milder downturn, the ratio of public spending to GDP was noticeably higher than only eight years earlier for both the Big Seven within the OECD (the United States, Japan, Germany, France, the United Kingdom, Italy, and Canada) and for the OECD's smaller industrial democracies (see figure 1).

FIGURE 1
TOTAL OUTLAYS OF GOVERNMENT AS A PERCENTAGE OF GDP IN THE G-7 COUNTRIES AND SMALLER OECD COUNTRIES, 1960–1992

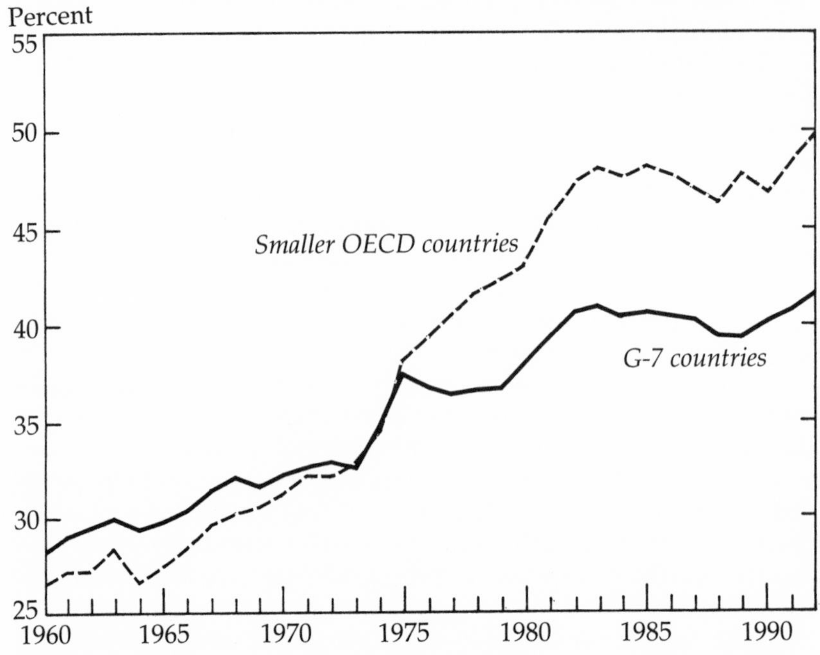

NOTE: G-7 = United States, Japan, Germany, France, the United Kingdom, Italy, and Canada.
SOURCES: For 1960–1989, Organization for Economic Cooperation and Development, *Historical Statistics* (Paris, OECD, 1992), table 6.5; for 1990–1992, derived from *OECD National Accounts 1980–1992* (Paris: OECD, 1994) and *OECD Economic Outlook*, no. 54, December 1993, p. 124

Furthermore, whatever the fate of the West's Left-leaning political parties, there is no good reason to presuppose that expansion of the state sector in modern societies should be contingent upon popular enthusiasm for statist or *dirigiste* ideologies per se. Growth of the state may be driven instead by an increased demand for services by those who view themselves as proper customers and consumers of what (in their view) government should provide. The modern voter and his agents have demonstrated already that they are quite capable of inveighing against the heavy burdens of an excessive government even as they militate for additional state subventions. As long as free individuals treat government as a superior good, there is no reason to expect its role in economic life to diminish over time.

As we look toward the future, it seems distinctly possible that we may be entering an era characterized by *both* the extension of liberal and accountable governance *and* the further expansion of the sway of the state. What would this presage for the tenor of daily life in our own nation and in the diverse societies of a world not yet arrived?

That question may be answered by posing another. Posit for a moment that the governments of this future world are constituted exclusively of constitutional democracies respectful of the rule of law. Presume that these democracies all choose to construct large and powerful administrative apparatuses that pursue activist, problem-solving policies. Assume finally that these policies are everywhere, without exception, implemented with only the highest of intentions and that in their formulation they somehow manage to escape the distortions that powerful claimants tend to impose on all big governments through preferential exactions.[15] What will befall people, in this idealized world of the future, when the meliorative actions of their own far-reaching states turn out to be animated by inaccurate data, to be informed by a misreading of available information, or simply to intrude into the wide reaches wherein numbers—though generated—cannot be expected to provide informed and reliable guidance? What, moreover, will the misdirection of well-intentioned international policies through the misuse of data portend for vulnerable groups in a world where tremendous differences in national incomes are likely to persist for generations to come—where millions upon millions of people will continue to live with very little margin for error?

It is in such an imaginary world of the future—in this vision that some would regard as the best of all possible worlds—that a tyranny of numbers will finally reign supreme. To imagine the workings of such a world, however, we do not have to dream of things yet to come. We are already exposed to misrule through numbers. As we shall see in the following chapters, not even the freest and most prosperous countries of our time can escape the toll that such misrule exacts. And in contemplating the character of this misrule, we gain a premonition of what lies in store in better times than we have known.

Misrule by the Numbers

The following chapters do not claim to offer a comprehensive overview of the misuse of statistics in government policy or even a systematic introduction to this phenomenon. This collection of studies instead touches on a variety of instances in which the policies of liberal and affluent states have been miscast or deleteriously directed through an ill-advised use of, or reliance on, statistical data. As we shall see, when a problem-solving government uses inaccurate data, selects accurate but inappropriate statistical indicators, or misanalyzes available facts and figures, the consequences are typically direct and injurious—not only for the citizens of the country concerned but at times for peoples in distant lands as well.

These chapters also suggest a deeper, even more troubling question. The question, simply put, is this: how much can the problem-solving state actually know, or learn, about the world that it is assigned to affect and actively improve?

If misrule by statistics were nothing more than a matter of avoidable errors by official analysts and policy makers, the solution would be to hire a better class of functionaries. If, on the one hand, such misrule derived principally from institutional tensions within the problem-solving state—selective use of figures, for example, by interest groups intent on championing their own policy preferences—the problem could be addressed as an issue of bureaucratic openness and accountability. If, on the other hand, there should be immutable limits to the knowledge that a state may acquire[16]—and if the problem-solving state should be disposed to trespass on those boundaries routinely—then an irresolvable problem would be lodged at the heart of the liberal, meliorative project. For the very state apparatus that was built on numbers would also be bound, by its design, to misuse numbers; and the system expressly dedicated to the protection of the individual and the practical promotion of the common good would necessarily inflict unintended injury on its intended beneficiaries. Under such circumstances, a tyranny of numbers would be both inescapable and seemingly unassailable—for enlightened and accountable governments would be enforcing it, and sovereign populations would be demanding it.

The Paradox of Poverty. Over the past generation, perhaps no problem in the United States has attracted so much federal attention, or absorbed so many public resources, as the paradox of persistent poverty within our increasingly affluent society. Between the late 1960s and the early 1990s, per capita output in the United States rose by more than half. Yet, over this same period, the officially measured poverty rate, instead of steadily declining, actually registered a *rise* in the proportion of the country subsisting on incomes below the established poverty line.

The persistence, even increase, in poverty rates is all the more troubling in light of the substantial and steadily increasing commitment of government funds expressly intended for its alleviation. By 1990, public expenditures in America on benefits for persons with limited income exceeded $200 billion—a sum that would average out to more than $6,000 for every person counted as poor in the country that year.[17]

In theory, such large amounts of money should have been far more than enough to eliminate officially defined poverty altogether: in 1990, after all, the poverty threshold for a family of three was set at under $10,600, and the total income deficit of all Americans below the poverty line was estimated to be less than $60 billion.[18]

For at least a decade, American scholars and policy makers have been engaged in extended and often contentious debate over the reasons why government social programs and antipoverty policies have so conspicuously failed to reduce the nation's poverty rate.[19] Yet for all the analytics and polemics that have been devoted to this question, one interpretation has rarely been offered: namely, that the famous poverty rate may not actually provide a good measure of material deprivation for contemporary America. Chapter 1 explores this proposition and concludes that the poverty rate is an arbitrary and, in some ways, a seriously misleading statistical indicator.

Over the years, critics have lodged a number of complaints about the poverty rate: that it neglects cost-of-living differences across the country; that it uses inappropriate deflators in its inflation adjustments;[20] that its poverty thresholds for different types and sizes of households are arbitrarily constructed;[21] that it does not properly value noncash benefits provided by public welfare programs. These are all valid criticisms; indeed, in recent years the Bureau of the Census, which compiles America's poverty data, has responded to some of these points by devising alternative poverty measures reflecting some of these considerations.

In a real sense, however, these criticisms miss the central problem. The problem is that *the poverty rate measures the wrong thing*. The poverty rate, after all, is estimated on the basis of reported annual household *income;* material deprivation, however, can be gauged only by patterns of *consumption*. By focusing on income rather than on consumption—which is to say, expenditures and purchasing power—our antipoverty policies have been guided by a false compass for thirty years.

Why did the American government select such a fundamentally flawed indicator of the incidence of poverty in the first place? The answer seems to be little more than bureaucratic expedience. In the early 1960s, when the poverty rate was devised, household income numbers were in hand and were collected annually; expenditure surveys, conversely, had been conducted irregularly, often with intervals of a decade or more between them. The staff and structure were in place in the early 1960s for an annual income-based poverty measure; an income-based poverty measure was correspondingly devised and formalized.[22] In the years since the poverty rate was first publicly presented,[23] the American government has made no serious effort to

FIGURE 2
CONSUMPTION VERSUS INCOME-BASED MEASURES OF POVERTY IN THE UNITED STATES, 1959–1989

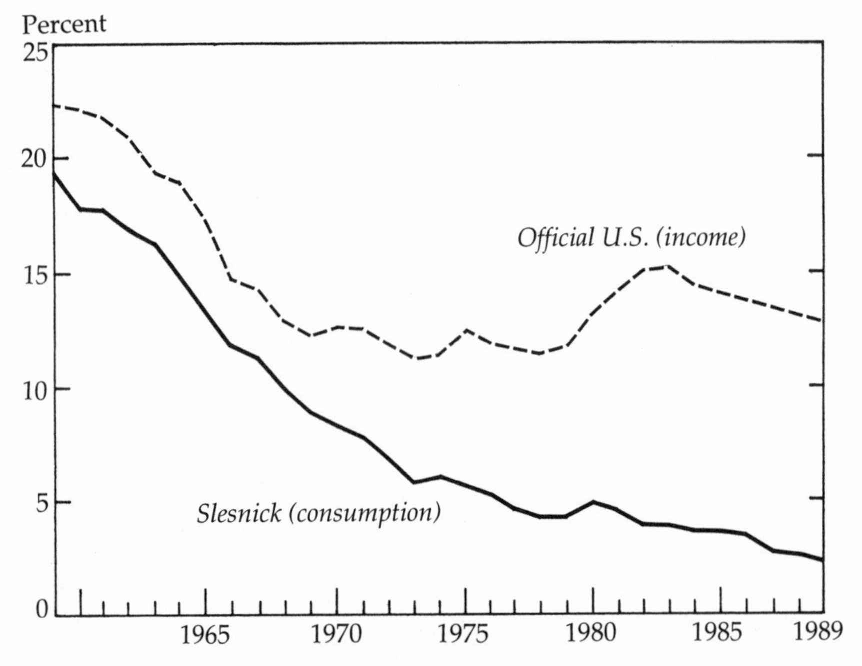

NOTE: Slesnick index is based on National Income and Product Accounts. For details and methodology, see source.
SOURCES: Daniel T. Slesnick, "Gaining Ground: Poverty in the Post-War United States," *Journal of Political Economy* (February 1993), pp. 1–38; and U.S. Bureau of the Census, *Poverty in the U.S. 1990* (Washington, D.C.: U.S. GPO, 1991), table 2, p. 16.

develop a more meaningful and intrinsically reliable index of material deprivation for our country.[24]

As it happens, a consumption-based index would present a dramatically different picture of material poverty in modern America from the one we have become accustomed to viewing. One recent attempt at a consumption-based measure (devised by Daniel T. Slesnick of the University of Texas) lays out the contrast (see figure 2). This index begins with the poverty thresholds originally stipulated by the U.S. government in 1965 but then adjusts these against changes in household spending patterns, also taking consumed noncash benefits into account.[25] These specific calculations depend on a number of assumptions, each of which might be contested in particular respects.

Whatever the adjustments, however, a consumption-based measure of deprivation would report unmistakable improvements in the U.S. poverty situation over the 1960s, 1970s, and 1980s. Such a reading, in fact, would be broadly consistent with the results reported in chapter 1.

With a clearer picture of the actual material circumstances of our country's vulnerable groups, a rather different policy debate might have been framed in the United States over the past decade. For while absolute deprivation, in a material sense, has been on the wane, dependence on government largesse among the public at large has been steadily on the rise. By 1990, more than 50 million Americans—more than a fifth of the nation—lived in households that accepted public means-tested benefits.[26] In that same year, more than half the country's black families were seeking, and obtaining, such resources. Among America's white families, the incidence of welfare benefit recipience was markedly lower; even so, more than a sixth lived in households on means-tested assistance. In 1990, indeed, more than a quarter of America's white families with children under eighteen years of age were enrolled in means-tested public assistance programs—a distinctly higher fraction than obtained among American *blacks* at the time the poverty rate was devised a quarter of a century earlier. Progress against material deprivation, it would seem, was indeed purchased over that generation—but at the price of the economic independence for large numbers of previously self-reliant Americans. Presented publicly, this would likely have been a matter of deep concern—but neither our citizenry nor our policy makers possessed the numbers that would have permitted them to discuss the trade-off.

Infant Mortality. The U.S. infant mortality situation is another example of a domestic problem in which policy prescriptions have been distorted by numbers—or, more specifically, by the absence of the sort of numbers that could help concerned citizens and policy makers better understand why so many babies are dying in America today. As I show in chapter 2, the conventional presumption in public health and public policy circles today is that our country's high level of infant mortality—and despite our national affluence, infant mortality in the United States is reported to be higher than in the great majority of Western, industrial democracies—can be explained by our unusually high prevalence of child poverty and by lack of access to health care under our current medical system. Both explanations are plausible on their face, but neither seems convincing on closer examination.

For reasons already mentioned, the poverty rate cannot provide an accurate depiction of comparative levels of material deprivation among the OECD countries—and there is little evidence to suggest that *consumption* levels for some large fraction of America's children are distinctively lower than in most other Western societies. As for medical care, international mortality data show that U.S. babies (black and white alike) are more likely to survive the critical perinatal period if born at highly vulnerable low birthweights than are babies from Japan or Scandinavia—places whose infant mortality rates today are considered exemplary. This differential strongly suggests that the American medical system is, in fact, making its resources generally available to the country's newborns—and that these are resources of high quality.

While many aspects of the American infant mortality puzzle remain unclear, one major reason the U.S. rate has been relatively high in recent decades seems to be that our society has been frighteningly efficient at turning out low-birthweight, high-risk babies. For American blacks, but also American whites, the incidence of low birthweight is high in comparison with other Western, industrialized societies. This distinction might prompt us to wonder about attitudes and practices of American parents-to-be that might expose their babies to risk. Surprisingly, despite the enormous data-gathering capabilities of our national vital and health statistics system, little information of this sort is available.

With respect to a mother's attitudes toward her pregnancy, for example, the U.S. government prepares but one survey—with only a few questions on this topic—and conducts it only every six years. Even so, results from the last round are intriguing—and arresting. As we see in chapter 2, the 1988 cycle of this National Survey of Family Growth suggests that a mother's attitude toward her pregnancy (that is, whether she considered it unwanted or mistimed at conception) is a much more powerful predictor of the risk of low birthweight for her baby than her income level.

If attitudes, and attendant behavior, play a role in explaining our contemporary infant mortality problem, it would seem reasonable to inquire about the health risks conferred on babies by their parents' lifestyles and marital patterns. Chapter 2 presents evidence that out-of-wedlock birth exposes the modern American baby to a heightened risk of infant death, even after the racial and social differences between married and single mothers are taken into account. The evidence marshaled to make that case, however, is less comprehensive than one would wish. The reason is straightforward: a long-standing lacuna of the U.S. system of health statistics involved the collection of information on the characteristics of out-of-wedlock births and deaths. During the 1960s, the 1970s, and on into the 1980s, the U.S. government simply did not collect the sort of data that would permit comparisons of infant mortality rates for births within and outside of a marital union.

If this was an understandable oversight when out-of-wedlock births accounted for 5 percent of the national total (as they did in 1960), it was less defensible by 1980, when the proportion of unmarried mothers was three and a half times higher. (By 1990, nearly 30 percent of the country's babies were born outside marriage.)

In the early 1980s, the American government began to ready a new compilation that would provide information on many risk factors facing out-of-wedlock babies—although this was not its primary purpose. The Linked Birth and Infant Death Project of the National Center for Health Statistics (NCHS) set about matching up birth certificates (which contain information about the mother's marital status) with death certificates for babies on a nationwide basis; by the late 1980s, these data, while mostly unpublished, were nevertheless in theory available. Even so, rather less was actually available than one might have supposed. Because of the idiosyncrasies of vital registration practices in many of the country's most populous states, for example, marital status for more than 40 percent of the nation's mothers was merely *inferred* as recently as 1988. The NCHS, for its part, seemed to be in no hurry to use its new, unpublished information to research the characteristics of health risk encountered by the infants of single mothers.

Why the reluctance to examine a topic that might help to explain the U.S. infant mortality problem—and possibly help save lives? As it happened, many researchers and policy makers seemed to feel that inquiries in this area were, so to speak, illegitimate. Some scholars explicitly challenged the motives of investigators who pursued the issue and raised the charge of blaming the victim.[27] Dramatics are by no means new to the academy or the public square when family matters come under discussion. Yet what is most noteworthy about this point of view, especially when it is maintained by public health professionals, is that it does not identify newborns exposed to heightened risk of death as the victims in the drama that it frames.

A closer look at the relationships between parental lifestyles and infant health risks might well spark vigorous public debate in the United States. It is possible that such a debate would revisit the questions of parental obligations, responsibilities, and duties; it is easy to imagine how it might assume a normative thrust. At the risk of presenting a value judgment, I would argue that there is nothing wrong whatever with such a use of statistics. Indeed, as Kenneth Prewitt contended during his tenure at the Social Science Research Council, statistical data can significantly contribute to the moral argumentation in which free peoples must engage if they are to determine national objectives and select means for achieving them.[28] The moral issues that confront a nation cannot be finessed by even the most expert manipulation of its numbers—and numbers bearing on the moral issues of the day can be avoided or ignored only at a cost ultimately measured in public well-being and the quality of political life.

Foreign Policies. As mentioned already, the misuse of numbers affects foreign as well as domestic policies. Chapters 3 through 6 may offer some indications of the impact of our misreading the performance of the Soviet-style system during our long postwar confrontation with Soviet-bloc communism.

In a sense, Soviet-style tyranny may be said to have been undone by statistics. Of all systems of government to date, the workings of the centrally planned economy are the most dependent on obtaining huge volumes of accurate data—yet the Soviet-style state is incapable of generating these. As a theoretical proposition, this problem was raised more than sixty years ago, when Ludwig von Mises, Friedrich von Hayek, and others made the case that a Socialist economic system could not produce the information it required for its own decision makers.[29] In a famous subsequent debate, Oskar Lange (later to be a member of the Polish State Council) countered that a centrally planned economic system could, in fact, approximate the information functions performed by the market-oriented economy.[30] Lange was wrong, as we now know.[31]

If its inability to generate the statistics needed for its own economic management was to establish a fundamentally unstable base for the Soviet project, the timing and particulars of its demise were due to a variety of other, more specific factors. For the most part, however, these factors and their significance were not generally recognized by Western students of the Soviet system, even in the years preceding its collapse.[32] Policy makers in the West seem typically to have succumbed to similar misassessments of the strength and the weaknesses of the Soviet system, even as that system approached its final crisis. To no small degree, these pervasive misassessments turned on the misanalysis of available data concerning the Soviet bloc's economies and societies.

Central to the Marxist-Leninists' claim to power, for example, was and is the assertion that their doctrine provided unique insights into the plight of the poor and unrivaled possibilities for improving their condition. This assertion was advertised relentlessly by every standing Communist state, insofar as it figured prominently in both their quests for domestic legitimacy and their international struggles against "imperialism." By the late twentieth century, this assertion had become so familiar that it was often accepted uncritically even by self-professed anti-Communists. Yet as chapter 3 attempts to demonstrate, a wide array of empirical data available before the end of the cold war not only challenged this assertion but effectively undermined it.

Intrinsic properties of the centrally planned economy—identified early on by its theoretical critics—made the measurement of its true performance an extraordinarily difficult task. Comparing the performance of the Soviet-style economy with that of a market-oriented economy presented fundamental, ultimately insoluble dilemmas, for there was no single, unambiguous method for the common valuation of their output. Under such circumstances, demographic numbers may be seen to have provided unusual apertures for social and economic insights. Population counts and vital statistics, after all, did not beg the measurement problems incumbent in the "people's economy": determining whether someone is alive or dead is considerably easier than determining the value of a Soviet sausage.

Chapters 4 and 5 attempt to show that population figures from the countries in question pointed to deep systemic troubles in both Eastern Europe and the USSR during their final generations under communism. From the mid-1960s through the 1980s, for example, the countries in which Red Army socialism held sway suffered stagnation, and even retrogression, in general health conditions. Long-term rises in death rates for industrialized countries not at war were historically unprecedented—yet now they were not only unique to the Warsaw Pact states but ultimately common to all its members. The health crisis that gripped these states during their last decades spoke directly to their living standards. But it also indicated constraints on their production possibilities. And insofar as the anomaly of rising death rates in a world characterized by generalized improvements in health may betoken regime fragility, this extended health crisis may have signaled the mounting pressures for systemic breakdown.[33] With only a few exceptions, however, researchers and policy makers in the West failed to appreciate the import of such demographic trends.

American policy toward the Soviet bloc during the cold war was informed by the assessments of the U.S. intelligence community, in particular the analyses of the Central Intelligence Agency. One of the CIA's principal assignments during the cold war was to evaluate the strength of our Soviet opponent—including its economic strength. As noted in chapter 6, the CIA's effort to describe the Soviet economy was probably the largest research project ever undertaken in the social sciences. Unfortunately, it was also a project whose output was seriously wanting in a number of important respects. As late as 1988, for example—less than a year before the revolutions in Eastern Europe—the CIA was placing per capita output in the Soviet Union at more than two-thirds the level of the Netherlands and estimating total Soviet output to equal the GDPs of Japan and Western Germany *combined*. The same approach prompted the CIA to rank East Germany's per capita output higher than the European Community's for 1987.[34]

This is not to minimize the theoretical and the practical challenges that loomed before the CIA's students of the Soviet economy. As noted, those were monumental—and would have been even without Moscow's strictly enforced policies of statistical secrecy. What I argue instead in chapter 6 is that the CIA's published, unclassified estimates on Soviet economic performance in the years before the end of the cold war are suggestive of methodological errors, inconsistencies, and biases. The methodological problems suggested by its unclassified estimates would be consistent with a broad overstatement of the Soviet economy's output and rate of growth and with an equally serious understatement of its true military burden (or ratio of military expenditures to national output). With the collapse of the Soviet state and the normalization of communications between the population of the former Soviet Union and the outside world, it has become apparent that the CIA's estimates erred in the directions I outlined.[35]

What were the consequences of the West's general misassessment of the Soviet bloc's social and economic performance? In a tautological sense, we do not know, for we can never know what would have transpired if Western citizens and policy makers had been in possession of more accurate analyses during the decades of the cold war. The final downfall of the Soviet system followed a path that was in no sense historically inevitable: Mikhail Gorbachev could have died in a plane crash instead of rising to general secretary of the Communist party of the Soviet Union; Erich Honecker could have taken a page out of Beijing's book and shot the demonstrators in Leipzig in 1989. With all the caution appropriate for contemplating the counterfactual, we may nevertheless wonder whether a better analysis of available data on the Soviet bloc might not have helped hasten the end of the cold war and contributed to an earlier release of those captive peoples from their bondage.

Defective Policies. With the end of the cold war, the issues addressed in chapters 7 through 11 promise to become more typical of the cadence of our country's foreign policy concerns. Promoting international economic development, redressing poverty in other countries, and attending to the world's food and population problems, after all, are today expressed American preferences—either on the part of the public at large or by our national leaders.

Poor populations are vulnerable populations; they are more likely to suffer severely from setbacks and unexpected shocks. As these chapters attempt to illustrate, some of the shocks and setbacks with which low-income populations in Asia, Africa, and Latin America must currently cope derive directly from the meliorative policies we apply to them. Often the defects of these policies involve a serious misuse of numbers.

With respect to foreign aid and development assistance (chapter 7), the number most easily obtained by policy makers in donor countries is the volume of public funds transmitted to recipient states. Such figures, however, can hardly be trusted to provide an accurate register of the actual *impact* of development assistance on local economies or societies. Economies characterized by long-term recipience of official development assistance also seem typically characterized by a number of troubling structural distortions and governmental practices. These have adverse implications for both the material well-being and the political liberties of the citizens in the countries in question. A more genuinely humanitarian approach to development assistance would concern itself more squarely with the demonstrable, and probable, consequences of its interventions. As I attempt to show, the data such an approach would require are already being generated in many low-income countries, although the margins of error on these data are often high.

As my study on poverty in South Africa may demonstrate (chapter 9), some governments in the modern world deliberately decided *not* to develop a statistical profile of the entire population—perhaps to spare themselves the moral and political pressures for action that would flow from such information. The apartheid era, however, is now over; even when it prevailed, its statistical system was anomalous. For most contemporary low-income countries, shortcomings in the national statistical system speak to incapability, not studied indifference. Whatever the reasons, however, the still rudimentary state of the official statistical systems in many contemporary low-income countries raises the possibility that interested parties may more or less invent their own numbers to serve their policy preferences. This issue is addressed in chapter 10, which examines the United Nations Children's Fund (UNICEF) depiction of the social consequences of the "third world debt crisis." A more careful and scrupulous study of available data, I argue, would have resulted in a less alarmist assessment of social trends in low-income countries during the 1980s. Where worrisome social trends could be identified, they may be seen to relate less to the international financial flows upon which UNICEF fixated than to practices of local governments that UNICEF seemed to ignore.

Chapters 8 and 11 discuss the world's food and population situations. In their approaches to both problems, as I attempt to demonstrate, both local and foreign policy makers have posed needless risks and hardships to large numbers of vulnerable people through a misuse of figures. A proclivity in various policy circles for exaggerating the prevalence of severe malnutrition in the modern world, I argue, has hampered international efforts to deal more effectively with extreme nutritional distress in those places where it does exist. By the same token, I argue, careless—or all-too-careful—use of population data has encouraged the mislabeling of policy-induced economic and social difficulties in Asia, Africa, and Latin America as "population problems," and in the process it has diverted some of the pressures that would otherwise militate for their reform. There is, I attempt to show, a major population problem looming in the future. Holding today's governments constant but projecting demographic growth forward, an ever larger share of humanity would be living under governments that do not uphold the rule of law or respect the rights of their own citizens. This prospect, however, highlights a *moral* problem near the center of modern international order—even if that moral problem cannot itself be couched in terms of numbers.

Moral Reasoning

Throughout the studies in this collection, cold statistics and purportedly value-neutral numbers seem to lead to moral issues in policy. Perhaps this should not surprise. When all is said and done, there can be no substitute for moral reasoning in human affairs, try as men may to devise one. The statistics-oriented, meliorative state may be new, but the question of how to use knowledge in a morally responsible manner is not.

When the novelty of the "probabilistic revolution"[36] has faded a bit, citizens around the world may recognize that policy-by-statistics is an approach to government based on a highly specific set of epistemological and philosophical assumptions. (Monarchism may sound antique today, but its epistemological and philosophical limitations were less self-evident a few centuries earlier.) Like all previous political arrangements, the meliorative, problem-solving state is intrinsically imperfect and imperfectible; its novel burden is the expectation and belief that an engine fueled by numbers will be capable of drawing the mass of humanity steadily closer to an earthly perfection. As the limits of this new form of governance come to be more generally recognized, we may hope that future generations will gradually learn to cope with its inherent flaws and its peculiar risks.

1

Economic and Material Poverty in Modern America

For many years, economists and policy makers have voiced concern about what is said to be an increasing number of poor people in the United States. Their attention has been focused for the most part on the rise in the measured rate of poverty since the early 1970s. According to the estimates of the U.S. Census Bureau, the fraction of the civilian, noninstitutionalized population living below the official poverty line underwent sustained increase during most of the 1970s and all of the early 1980s. Between 1973 and 1983 the "poverty rate" rose from 11.1 percent of the population to 15.2 percent, or by well over a third. Although the poverty rate declined somewhat during the subsequent years, it was still placed at 13.6 percent in 1986—higher, both relatively and absolutely, than it had been a decade and a half earlier.

Yet strangely, while the measured rate of poverty was rapidly rising, other indicators of economic well-being—growth of disposable income, improvements in educational qualifications, advances in health—suggest that the 1970s and early 1980s were a period of general progress toward greater affluence and prosperity. During this period the poverty rate seemed to become detached from those economic forces that traditionally affected most Americans' financial well-being.

In the 1960s and early 1970s, for example, an increase in per capita GNP of 3 to 4 percent corresponded with a decline in the measured poverty rate of about one point; yet in the late 1970s and the early 1980s a similar increase in per capita GNP corresponded with a slight *rise* in the poverty rate. Nor did changes in the unemployment rate seem adequate to the task of reducing the new poverty. Between the mid-1960s and the early 1970s, during the heyday of America's postwar boom and the War on Poverty, any given rate of unemployment, if held constant, was consistent with a drop in the measured rate of poverty of almost one point a year. By the late 1970s and early 1980s a given rate of unemployment was consistent with an annual *increase* in the poverty rate of nearly one-third of a point. This rise in measured poverty, moreover, was occurring when the U.S. government, through its public policies, was devoting ever larger absolute amounts of resources to the alleviation of material need (and for much of this period a larger *relative* fraction of national resources as well).

If these figures raise many questions about the well-being of the American populace, they may also suggest that some of the economic indexes commonly used today to depict the degree of poverty in the United States provide little insight into the physical and material well-being of the poor. In other words *economic* poverty (which tells us about the income level of the poor) may be quite different from, and arguably less important than, what can be called *material* poverty (which tells us about their health and welfare). Economic data alone may even provide a seriously misleading impression of the physical circumstances of given groups over time.[1]

By contrast, traditional indicators of *deprivation* such as health and nutrition might give us a less ambiguous and more consistent idea of how the poor—whatever their numbers—are doing. Important in and of themselves for their broad correspondence with what is generally recognized as "poverty," health and nutrition figures are also, in meaningful ways, less subject to error than economic data. It is easier, after all, to determine whether a person is alive or dead than to determine income—and if anything more important. Indeed, information about life expectancy, infant mortality, suicides, birth weights, undernutrition, and diet may tell more about the material condition of Americans in poverty than conventionally quoted statistics about income do.

Health and Poverty

Good health comes close to being a universally desired personal attribute. Although important for reasons entirely unrelated to money, it also has a real economic value. This may be appraised in a variety of ways. Thus Dan Usher's book *The Measurement of Economic Growth* attempts to estimate the utility to consumers of their own increased consumption when their lives lengthen. On the basis of some arbitrary but not implausible assumptions, Usher concludes that an imputation for improved life expectancy would, for the years 1930–1974, increase the measured rate of growth in U.S. GNP per capita by something like 25 to 29 percent. Such imputed benefits from improved health, though by no means trivial to the individuals affected, are never directly registered in national accounts.

Yet poor health still has direct and indirect economic costs that can be measured more precisely. The National Center for Health Statistics (NCHS, now part of the Centers for Disease Control) has attempted to do just this for the year 1980 in the United States. The NCHS found that those families with the lowest measured income had the highest direct medical costs per capita (although these families did not necessarily assume the costs themselves). Based on several assumptions about lifetime income, the NCHS also estimated the indirect costs of poor health that would result from earnings lost because of illness. Earnings lost because of such indirect health costs were found not only to be relatively higher for poor families than for families as a whole but also to be higher in absolute terms.[2]

The significance of such income losses should not be underestimated. If illness among poor and low-income families could be reduced, the effects on the measured rate of poverty could be consequential. To take a hypothetical example, reducing the poverty population's annual per capita income losses because of poor health so that they equaled the national average would reduce the measured rate of poverty in the population as a whole by about one-fifth.

Such speculation about remaining health problems, however, should not obscure the actual health advances in recent decades. After the rapid rises in life expectancy in the 1940s, improvements in life expectancy in the 1950s and early 1960s were slow (table 1–1). Improvements in life expectancy at birth began to accelerate again in the mid-1960s. Over the decade 1973–1983 they were three times as great for the country as a whole as during the 1960s. Improvements were particularly sizable for the nonwhite population, which historically has had lower life expectancies. Over the course of the 1960s, the gap in life expectancy at birth between whites and nonwhites narrowed by only about seven months. Yet between 1973 and 1983 that gap closed by 2.2 years. Improvements in life expectancy between 1973 and 1983 were greater for nonwhites than for whites not only at birth but also in youth and middle age.

Infant Mortality

The infant mortality rate plays a major role in determining overall life expectancy; it is also widely regarded as an indicator with broader reference to social development. As seen in table 1–2, during the late

TABLE 1–1
EXPECTATION OF LIFE AT BIRTH, 1940–1983

Year	*All Persons*	*Males*	*Females*	*Whites*	*Nonwhites*
1983[a]	74.7	71.0	78.3	75.2	71.3
1980	73.7	70.0	77.5	74.4	69.5
1973	71.4	67.6	75.3	72.2	66.1
1970	70.8	67.1	74.7	71.7	65.3
1960	69.7	66.6	73.1	70.6	63.6
1950	68.2	65.6	71.1	69.1	60.8
1940	62.9	60.8	65.2	64.2	53.1
Change in Life Expectancy (Years)					
1973–1983	3.3	3.4	3.0	3.0	5.2
1960–1970	1.1	0.5	1.6	1.1	1.7
1955–1965	0.6	0.1	0.9	0.5	0.4
1940–1950	5.3	4.8	5.9	4.9	7.7
Change in Life Expectancy, Nonwhite minus White (Years)					
	All	*Males*	*Females*		
1973–1983	2.2	2.0	2.3		
1960–1970	0.6	–0.4	1.6		
1955–1965	–0.1	–0.5	0.3		
1940–1950	2.8	4.2	2.4		

a. Preliminary figures.
SOURCE: U.S. Census Bureau, *Statistical Abstract of the United States* (Washington, D.C.: U.S. Department of Commerce, various years).

1950s and the early 1960s there was comparatively little progress against infant mortality in the United States. Infant mortality resumed a more rapid decline in the mid-1960s and, like the increase in life expectancy, was particularly rapid between 1973 and 1983. In absolute terms, infant mortality fell further for nonwhites and blacks than for whites between 1965 and 1983, although it remained significantly higher than the white infant mortality rate. One further indication of health progress comes from the Indian Health Service, which is charged with providing public medical care to Native American populations on and around Indian territories. If its figures are correct, infant mortality rates for the areas serviced were lower than the rate for the country as a whole in 1982. As recently as 1955 they were measured to be nearly two and a half times the national average.

The correlation between infant mortality rates and official poverty rates for children became progressively less close between 1959 and 1983. Table 1–3 shows that until 1973 the poverty rate for children

TABLE 1–2
INFANT MORTALITY RATES, 1940–1983
(rate per 1,000 live births)

Year	*All Races*	*White*	*Nonwhite*	*Black*	*Ratio of Nonwhite to White (white = 100)*
1983	11.0	9.7	16.8	19.2	173
1980	12.6	11.0	19.1	21.8	174
1973	17.7	15.8	26.2	28.1	166
1970	20.0	17.8	30.9	32.6	174
1965	24.7	21.5	40.3	41.7	187
1960	26.0	22.9	43.2	44.3	189
1955	26.4	23.6	42.8	NA	182
1950	29.2	26.8	44.5	43.9	166
1940	47.0	43.2	73.8	72.9	171
		Percent Change			
1973–1983	–37.3	–38.6	–35.9	–31.7	
1960–1970	–23.1	–22.3	–28.5	–26.4	
1955–1965	–6.4	–8.9	–5.8	–5.0[a]	
1940–1950	–37.9	–38.0	–39.7	–39.8	

NA = not available.
a. This figure is for the years 1950–1965.
SOURCES: U.S. Bureau of the Census, *Statistical Abstract of the United States: 1986* (Washington, D.C.: U.S. Department of Commerce, 1986); "Advance Report of Final Mortality Statistics, 1983," *Monthly Vital Statistics Report,* vol. 34, no. 6, supp. 2 (September 26, 1985); National Center for Health Statistics, *Vital Statistics of the United States* (Washington, D.C.: U.S. Department of Health and Human Services, various years).

roughly paralleled the infant mortality rate. But after 1973 this correspondence completely broke down. Between 1973 and 1983, for example, the measured rate of poverty for black children rose by more than one-eighth, yet the infant mortality rate for black Americans fell by nearly a third. During the decade beginning in 1973, in fact, child poverty rates and infant mortality rates were moving in opposite directions for children of all races.

Infant mortality rates are affected by a number of different proximate factors, one of the most important being low birthweight. Low-birthweight babies (defined as those born under 2,500 grams—about 5.5 pounds) have much higher mortality rates than those who have higher weights at birth. Mortality risks by weight are, of course,

TABLE 1–3
INFANT MORTALITY RATES PER 1,000 LIVE BIRTHS AND ESTIMATED POVERTY RATES IN PERCENTAGES FOR RELATED CHILDREN UNDER EIGHTEEN, 1960–1983

Characteristic	*1960*	*1965*	*1973*	*1975*	*1980*	*1983*
All races						
Infant mortality rate	26.0	24.7	17.7	16.1	12.6	11.0
Poverty rate	26.5	16.3	14.2	16.8	17.9	21.7
Whites						
Infant mortality rate	22.9	21.5	15.8	14.2	11.0	9.7
Poverty rate	20.0	14.4	9.7	12.5	13.4	16.9
Nonwhites						
Infant mortality rate	43.2	40.3	26.2	24.2	19.1	16.8
Poverty rate	66.6	59.0	38.9	38.0	38.3	42.8
Blacks						
Infant mortality rate	44.3	41.7	28.1	26.2	21.8	19.2
Poverty rate	NA	47.4[a]	40.7	41.4	42.1	46.3

NOTES: "Related children" are children brought up in families (as opposed to orphanages).
NA = not available.
a. This figure is for 1967.
SOURCES: U.S. Bureau of the Census, Series P-60, various years, and *Statistical Abstract of the United States*, various years.

on a continuum, but the odds of dying in infancy are today about twenty times as great if one is born below this low-birthweight line as if one is born above it. According to NCHS data, the overall fraction of lower-weight births in the United States has remained relatively stable since 1950, while the infant mortality rate has consistently declined.

The U.S. infant mortality rate is frequently compared unfavorably with those of other developed Western nations. But this is not altogether correct. (For a fuller treatment of this argument, see chapter 2.) In 1982, according to the U.S. Public Health Service, the United States appeared to be almost in the middle of the range for such countries.[3] The American infant mortality rate for that year was lower than those of such places as Austria, West Germany, and New Zealand but higher than the rates reported in Scandinavia, Switzerland, and Japan. The low-birthweight ratios of these latter nations, however, are significantly below that of the United States. In relation to its low-birth-weight ratio of 1980—which was the second highest among industrialized nations—the United States actually has an unusually low rate of infant mortality.[4]

TABLE 1–4
INFANT MORTALITY RATES FOR LEGITIMATE AND ILLEGITIMATE BIRTHS, 1960
(per 1,000 live births)

	Age of Mother				
Characteristic	15–19	20–24	25–29	30+	All ages
White births	28.1	21.4	20.0	22.3	22.2
Illegitimate	33.0	30.3	32.7	39.0	33.0
Legitimate	27.7	21.2	19.9	22.1[a]	21.9
Nonwhite births	49.5	40.2	37.3	39.8	41.4
Illegitimate	52.7	48.5	48.9	52.0	51.2
Legitimate	47.2	38.2	35.4	40.8[a]	38.3

a. Estimates derived from thirty- to thirty-four-year-olds only.
SOURCES: National Center for Health Statistics, *A Study of Infant Mortality from Linked Records by Age of Mother, Total Birth Order, and Other Variables,* series 20, no. 14 (Washington, D.C.: U.S. Department of Health, Education, and Welfare, September 1973); NCHS, *Vital Statistics of the United States: 1960,* vol. 1 (Washington, D.C.: U.S. Department of Health, Education, and Welfare, 1963).

Another proximate risk to infant health is illegitimacy. The U.S. government gathers data on illegitimate births but did not customarily collect data on illegitimate deaths. When it did so in 1960, it found that infant mortality was one-third higher for illegitimate nonwhite babies than for legitimate ones, and more than 50 percent higher among whites (see table 1–4). Infant-mortality rates, moreover, were higher among illegitimate babies not only by race but also by maternal age. Given this apparent correlation between illegitimacy and high infant-mortality rates, one might have expected an increase in the latter from 1950 to 1983, when illegitimacy was undergoing a much-publicized increase. But what we find is exactly the opposite: infant mortality rates have declined while illegitimacy rates have soared. This decline seems to have occurred not because of the breakdown of the two-parent family, but *despite* it.

Infant mortality is not the only sort of death rate that provides insight into a nation's well-being. Some indication of self-assessed well-being may also be gleaned from the incidence of suicides. Suicide rates, although perhaps the reflection of a variety of factors, should nonetheless tell us something about the quality of life in the country. The rates rose sharply in the 1920s—a period of prosperity—and slumped just as sharply in the 1930s—during the depression. Although today suicide rates are much lower than in the earlier part of this century, they have been rising since 1950. Since 1960, suicide rates for men have been higher than for women, and sex-specific suicide rates have been higher for whites than blacks. At first glance, and perhaps paradoxically, suicide rates seem to be inversely correlated with measured poverty.

Health and Welfare

Where, and for whom, are health conditions in the United States poorest? Even after the progress of recent decades, this question can still be answered entirely without reference to the white population. To be sure, health conditions for nonwhites are not always worse than for whites. The state with the highest life expectancy in 1980, the latest period for which state life tables were computed, was Hawaii; it is predominantly nonwhite. Yet on a national basis life expectancy for nonwhites is significantly lower than for whites, principally because of the gap between whites and blacks. As of 1980, there was no overlap between white and black life expectancies in the fifty states and the District of Columbia; the highest black life expectancy in any state (Massachusetts) is lower than the lowest white life expectancy in any state (Nevada). In both 1970 and 1980 life expectancy for nonwhites was lowest in four southern states and the District of Columbia. With 1980 data, if blacks are separated from other nonwhite populations, one northern state joins the list: Illinois. Of all places in the nation, the District of Columbia has the lowest measured life expectancy at birth for blacks.

One may speculate that social welfare programs could have two opposite effects on the health of the American population. The obvious one would be to improve health directly and unambiguously by providing the public and the poor with better medical care, education, housing, and nutrition than they would otherwise have. But a countervailing effect might arise through the impact of social welfare programs upon the family—insofar as they subsidize those family patterns that are correlated with greater risks to infant survival, such as illegitimacy.

Further research must be done before this speculation can be rigorously tested. Data collected by the government of the District of Columbia, however, are intriguing (table 1–5). The municipal government in Washington, D.C., has constructed what it calls "a dependency indicator," a composite index reflecting the ratio of recipients of different means-tested public benefit programs to total population by census tract. Census tracts are then ranked, not according to poverty rates or income levels, but instead by their dependency indicator; infant mortality rises with that indicator as well, although the correlation is not

TABLE 1–5
HEALTH CHARACTERISTICS OF INFANTS IN WELFARE FAMILIES IN THE DISTRICT OF COLUMBIA, 1983

	Dependency Indicator				
Characteristic	City Average	Best	Moderate	Poor	Worst
Infant mortality rate (per 1,000 live births)	18.2	12.2	20.5	18.8	24.0
Neonatal mortality[a] (per 1,000 live births)	13.2	10.8	14.7	14.3	13.9
Postneonatal mortality[b] (per 1,000 live births)	4.9	1.4	5.8	4.5	10.1
Births to teenaged mothers (percent)	18.8	7.3	16.8	22.6	28.1
Out-of-wedlock births (percent)	57.4	29.3	55.4	68.2	75.5
Low-birthweight births (percent of live births under 2,500 grams)	13.1	8.9	13.2	14.3	16.2

a. "Neonatal mortality" refers to deaths in the first four weeks of life.
b. "Postneonatal mortality" refers to deaths in the first year of life but after the first four weeks.
SOURCES: District of Columbia Department of Human Services, *Vital Statistics of the District of Columbia: 1983* (Washington, D.C.: DHS Office of Policy and Planning, n.d.).

so tight. Income levels, measured poverty status, and the receipt of public benefits are all closely correlated. It would be interesting to attempt to control for income or measured poverty status and to see its effect on these indicators of child well-being in relation to welfare for children.

Nutritional Well-Being

The nutritional well-being of a population is more difficult to assess than is commonly understood. Short of clinical examination—a procedure hardly feasible for a nation as a whole—it is difficult to obtain the biochemical and metabolic data upon which to make confident pronouncements about nutritional status. Even with these, there remains the problem of relating deviations from specified ideals to their practical consequences in daily life. Compounding these difficulties is the nature of the appetite mechanism. Hunger is a subjective condition.

TABLE 1–6
EXPENDITURES ON FOOD AND BEVERAGES FOR FAMILIES OF CITY WAGE AND CLERICAL WORKERS, 1888–1961

Consumption Year	*Food and Beverage Expenditures as % of Money Income after Taxes*	*Food and Beverage Expenditures as % of Personal Expenditures*
1888–91	44.5	47.2
1901	49.7	52.4
1917–19	35.5	39.4
1934–36	38.7	40.2
1950	33.3	32.8
1960–61	26.6	28.2

SOURCE: Bureau of Labor Statistics, *Handbook of Labor Statistics: 1973* (Washington, D.C.: U.S. Department of Labor, n.d.), p. 325.

Persons who are clinically obese may feel constantly hungry, just as a person dying of starvation may have no will to eat.

But nutritional well-being may be gleaned from purchasing patterns of households—for example, the proportion of a household budget spent on food. Patterns of expenditure should in a very real sense reveal something about self-assessment of nutritional status.

Table 1–6 shows the proportion of expenditures on food and drink in relation to income and total expenditures for urban blue- and white-collar workers in surveys from the end of the nineteenth century to the middle of the twentieth century in America. Although there were fluctuations—a rise, for example, during the depression—and doubtless measurement errors, the trend was downward. The trend of decreasing proportional expenditures on food coincided with—and is consistent with—a rise in life expectancy.

In 1977–1978, at the time of the decennial National Food Consumption Survey of the Department of Agriculture (USDA), about 18 percent of the personal consumption expenditures of U.S. households, on average, went for food. For "low-income" households—in the USDA survey, those that could qualify for food stamps—the fraction appears to have been about 19 percent. The percentage of expenditures allocated to food by the "low-income" population in the United States was thus lower than for any Western European population for which comparable data were available. In fact, according to USDA analyses, the marginal propensity of the U.S. "low-income" population to spend money on food was nine to twelve cents on every additional dollar of income.

Every food stamp "dollar" was estimated to have increased expenditures on food by about twenty-five cents.[5]

By 1984, households in the lowest income quintile in the Bureau of Labor Statistics (BLS) Consumer Expenditure Study were allocating less than 19 percent of their personal consumption expenditures (PCE) to food. This contrasts with the original presumption in constructing the "poverty line" back in 1964 that a family in poverty in the United States would be spending roughly a third of its after-tax income on food. By 1982–1983, according to the BLS survey, even groups traditionally considered vulnerable to adverse circumstances were devoting small fractions of PCE to food. At a time when the measured poverty rate was rising for the black population, the share of food in PCE for the black population fell below 18 percent. Households headed by persons sixty-five or older, but with less than $5,000 reported annual income—virtually all of which would have been below the 1983 poverty line—devoted less than 20 percent of their PCE to food. For households headed by a person twenty-five or younger reporting an annual income of $5,000 or less—100 percent of which would have been below the 1983 poverty line—the share of food in PCE was about 21 percent.

Food budgets can be instructive in telling us not only how much is being spent but also what is being purchased. One would ordinarily expect persons feeling nutritional deprivation to attempt to stretch their food budget—to maximize the nutrients per dollar of food purchased. Results from the 1977–1978 National Food Consumption Survey suggest that lower-income households, for whatever reasons, purchased fewer calories with their food dollar than did low-income households without food stamps.[6] This result is quite in keeping with economic theory, insofar as food stamps reduce the relative price of all foodstuffs and afford recipients greater scope for the purchase of preferred items, whether or not these are nutritious.

The late M. K. Bennett, long-time director of the Stanford Food Research Institute, argued that deprivation could be adduced from the proportion of starchy staples in a diet. Between 1965–1966 and 1977, according to the USDA's National Food Consumption Survey, the fraction of inferior calories (cereals, tubers, and legumes) fell from 29.7 to 27.4 percent. ("Inferior" refers not to the nutritive value of such foods, but rather to their rankings in consumer preference.) Interestingly enough, the group with the highest estimated per capita caloric availability—the rural farm population—also had the highest share of "inferior" calories in its diet. The quality of the diet between surveys apparently rose, even as measured caloric availability per capita fell.

Recommended dietary allowances (RDAs) are frequently used as a basis for assessing the prevalence of malnutrition in a population.

TABLE 1–7
HOUSEHOLDS RECEIVING LESS THAN 100 PERCENT OF RECOMMENDED DIETARY ALLOWANCE, SPRING 1977
(percent)

	Percent of Households			
Money Income before Taxes, 1976	Total	Central City	Suburban	Non-metropolitan
Under $5,000	23.3	27.1	26.7	22.0
$5,000–9,999	25.5	25.6	26.1	24.8
$10,000–19,999	24.3	27.4	22.4	23.0
$20,000 and over	18.4	23.5	17.3	15.6
All households	23.3	26.1	21.8	22.2

NOTE: Recommended dietary allowance of caloric energy was determined in 1974.
SOURCE: U.S. Department of Agriculture, *Dietary Levels: Households in the United States, Spring 1977*, NFCS 1977–1978 Report no. H-11 (Washington, D.C.: Human Nutrition Information Service, March 1985), pp. 70–71.

Unfortunately the method of most such assessments is problematic. Even if accurate, RDAs provide an *average* recommendation. But needs vary—in fact, something like half of any population will need less than the average requirement *by definition*. Failure to appreciate this point can lead to serious problems of data interpretation. In table 1–7, for example, more than a sixth of all households with 1976 incomes above $20,000—the equivalent of almost $40,000 today—are estimated to receive less than their RDAs. Such a figure calls into question the meaning of all other estimated RDA shortfalls.

Physical Data on Nutrition

Nutritional deprivation may also be assessed by physical and anthropometric measures. One of these indicators is the age of menarche for girls. Research has shown that the onset of menarche is affected by the ratio of fat to total body weight and that below a certain threshold menarche may be delayed or even prevented. For a sample of women born between 1926 and 1944 and surveyed by the National Center for Health Statistics, nutrition apparently was better for urban than for rural girls; the difference in age at menarche was about four months. In this sample white urban girls reached menarche about half a year sooner on average than black rural girls. By the same token years of schooling—possibly a proxy for other family circumstances—coincide

TABLE 1–8
ESTIMATED DAILY CALORIC INTAKE OF U.S. POPULATION AND OF POVERTY-LEVEL POPULATION, 1976–1980

	Percentile Distribution		
Characteristic	Mean	50th	5th
All males, 6 months–74 years	2,381	2,187	976
Poverty males, 6 months–74 years	2,214	2,008	771
All females, 6 months–74 years	1,578	1,493	683
Poverty females, 6 months–74 years	1,557	1,460	606
Males, 1–2 years	1,331	1,259	653
Poverty males, 1–2 years	1,296	1,234	602
Females, 1–2 years	1,262	1,170	649
Poverty females, 1–2 years	1,359	1,298	632
Males, 15–17 years	2,817	2,629	1,133
Poverty males, 15–17 years	2,547	2,303	906
Females, 15–17 years	1,731	1,636	710
Poverty females, 15–17 years	1,771	1,650	765
Males, 25–34 years	2,734	2,577	1,213
Poverty males, 25–34 years	2,573	2,302	1,090
Females, 25–34 years	1,643	1,547	677
Poverty females, 25–34 years	1,711	1,575	621
Males, 65–74 years	1,829	1,723	841
Poverty males, 65–74 years	1,602	1,495	667
Females, 65–74 years	1,295	1,221	607
Poverty females, 65–74 years	1,223	1,163	464

SOURCE: National Center for Health Statistics, *Dietary Intake Source Data: United States, 1976–80,* series 11, no. 231 (Washington, D.C.: U.S. Department of Health and Human Services, March 1983).

with differences in age at menarche. Almost half a year separated girls with fewer than five years of schooling from those with some college or more. Age at menarche appears to have been dropping with each generation; the pace of decline may have been more rapid in recent generations for blacks than for whites, but this is not certain.

Table 1–8 provides further physical data on nutrition: measured caloric intake from a large-scale survey in the late 1970s. Actual measured caloric *intake* in the 1976–1980 survey was about 33 percent lower than measured caloric *availability* in the 1977–1978 USDA survey. (Caloric availability is the food available to consumers at supermarkets and other food-supply sources after transportation and storage losses.) The level of daily per capita intake indicated for Americans is about the same as the United Nations Food and Agriculture Organization estimates of available food supply in Bangladesh—which should caution us as to the worth of comparisons using these sorts of numbers. As may be seen, mean and median per capita caloric intakes for females below the poverty line were seldom lower than for females of all incomes; often caloric intake for poverty females was measured to be greater.

A final indication of nutritional status may be obtained from NCHS estimates of obesity in the United States based on anthropometric surveys in 1960–1962, 1971–1972, and 1976–1980. The measure used is probabilistic rather than clinical, but it is applied standardly over time. By this measure the proportion of adults twenty-five–seventy-four designated as obese rose slightly more for black men than for white men, and more for black women than for white women; a greater fraction of adult blacks of both sexes was designated as obese in the late 1970s than was true of whites. Among persons sixty-five–seventy-four—a group often feared to be vulnerable to nutritional deprivation—nearly a third were measured to be obese, a higher fraction than for any other adult age group of Americans in the late 1970s. More than 60 percent of black women in this age group were estimated to be obese.

Survey data, in short, do not seem to point to any large pockets of nutritional deprivation in modern America. Does this mean that there is no nutritional deprivation in modern America? Not necessarily; it may only mean that sophisticated large-scale surveys cannot locate nutritional deprivation in America today. Yet despite the increasing efficacy of governmental enumerative efforts, a certain small fraction of the population is known to slip through the statistical cracks. In 1980, in the most expensive and careful census to its time, an estimated net of 1 percent of the total population was missed. The population undercounted is thought to have been disproportionately black, male, and of working age (twenty-five–fifty-nine).[7] If the undercount provides the hard core of unidentifiables, the characteristics of households without telephones may indicate the characteristics of underenumeration in less-extensive and less carefully conducted surveys: disproportionately nonwhite, poor, and in households headed by someone young (table 1–9).

TABLE 1–9
PERSONS WITHOUT TELEPHONES, NOVEMBER 1983
(percent)

Category	*All Races*	*White*	*Black*	*Hispanic Origin*
All persons	7.2	5.9	17.3	16.4
Householder employed	5.9	5.0	14.3	13.7
Householder unemployed	17.5	15.2	25.4	23.4
Householder not in labor force	7.9	6.2	19.2	19.6
All households	8.6	6.9	21.2	18.3
1 person	11.4	9.3	28.8	26.2
2–3 persons	6.7	5.5	17.5	19.3
4–5 persons	7.6	6.4	16.9	16.6
6+ persons	13.4	9.5	25.5	19.0
All families	7.2	6.9	21.2	19.3
Total income under $5,000	20.3	24.3	37.3	41.7
$10,000–12,499	10.3	8.8	18.0	23.2
$20,000–24,999	3.1	2.6	8.8	6.9
$30,000–34,999	1.2	1.0	4.9	2.3
$50,000–74,999	0.6	0.5	1.5	0.4
All households	8.6	6.9	21.2	19.3
Households 16–24 years old	23.4	19.8	50.1	35.1
Households 25–54	8.5	6.6	21.3	18.2
Households 55–64	5.0	3.9	13.7	10.7
Households 65–69	4.5	3.5	12.5	9.3
Households 70+	4.6	4.0	8.9	14.5

SOURCE: U.S. Bureau of the Census, unpublished data.

Prosperous Paupers

In a sense, social policy may have reached its limits in the quest to "end hunger" in modern America. Those who can be identified do not seem to suffer pronounced nutritional deprivation; those who may suffer nutritional deprivation are not (perhaps cannot be) identified—and therefore cannot be helped.

What can be said about economic and material poverty in the United States today? A numer of points may be emphasized. First, the poverty index, by its method of construction, probably has been biased upward, and may have been biased increasingly upward over time. Between 1973 and 1983 it actually seemed to move in the opposite direction from some important physical indicators of well-being, such as death rates. Second, changes in family structure in the United States—and in particular the rapid rise in illegitimacy—have disturbing implications for the well-being of children, who are perhaps the country's most vulnerable group; physical measures of poverty—including mortality rates—capture and reflect some of these risks. Third, judging by material measures of health and nutrition, if more poverty is said to have existed in the mid-1980s than in the early 1970s, poverty today incontestably is less physically harsh and punishing. The condition of the poor in this sense has definitely improved.

But the results of American social policy over the past generation, for all the importance of these physical successes, do not immediately excuse the government, or the public that sustains it, from the criticism that in eliminating physical poverty we have created in our midst a class of prosperous paupers. The term "pauperism" is lost on the modern ear, but its meaning, as given by the *Oxford English Dictionary,* is specific: "dependence upon public relief, as an established condition or fact among a people."

There is less physical poverty in the United States today than a generation ago. Yet just as certainly there is, in some meaningful sense, considerably more pauperism in the nation. The quest to end poverty in the postwar era has been the scene simultaneously of material triumph and of social—one might even say moral—failure. This dilemma must inform any assessment of America's progress in its fight against poverty over the past generation and should be squarely faced in discussions about future U.S. social policy.

2

The U.S. Infant Mortality Problem in an International Perspective

The infant mortality rate—the number of children to die in their first year of life out of every thousand born—is widely viewed as an especially meaningful social indicator. It provides an immediate representation of the health situation for babies within any population, a matter of considerable importance in its own right. The infant mortality rate has also come to be viewed as a more general reflection of a society's capacity, or willingness, to protect its most vulnerable elements and thus in a sense as a measure of social development itself.

Over the course of the 1980s the U.S. infant mortality rate became a focus of increasing attention, commentary, and public concern. By comparison with many other countries, the U.S. infant mortality rate has not only been unusually high but has been lagging further and further behind the front-runners. The issue of the annual statistical compendium published by the U.S. Department of Health and Human Services in March 1990, *Health USA,* provides an official assessment of the problem.[1] In 1986—the most recent year for its international comparisons—the U.S. infant mortality rate is reported to have been higher than in twenty-one other countries or territories, including practically all of Western Europe, Hong Kong, Singapore, and East Germany. Between 1981 and 1986, despite a decline in its reported infant mortality rate, America's international ranking is shown to have fallen by three places. By 1986, according to these numbers, the U.S. infant mortality rate was more than 30 percent higher than Norway's, 50 percent higher than the Netherlands', and fully twice as high as Japan's.

America's comparatively high rate of infant mortality and its relatively poor performance in reducing its level are conventionally attributed to two underlying factors: the prevalence of poverty and a corresponding lack of adequate health care. Given America's general affluence—careful estimates suggest that price-standardized per capita output in the United States in 1987 was more than 30 percent higher than in Norway, Holland, or Japan[2]—the country's infant mortality problem is widely viewed as reflecting serious shortcomings and inadequacies in official income support and health care policies.

That view is accepted not only by many specialists in public health but also by a number of policy makers in Washington. In March 1990, for example, the chairman of the House Select Committee on Children, Youth, and Families opened a day of recent hearings on "Child Health: Lessons from Developed Nations" with the observation that

> childhood poverty, the greatest predictor of poor child health outcomes, is worse in the U.S. than in most other industrialized countries, and financial barriers are by far the most common and significant reason that women and children don't receive the health care that they need. By contrast, in Europe or Canada, no pregnant woman has to ask how, or where, she will receive prenatal care, or who will pay for it.[3]

He suggested moreover that "the time had come to thoughtfully consider the practices of other comparable countries, which in many important areas, are achieving better health and economic outcomes for their children and families, despite their smaller gross national product."[4]

Compelling though such interpretations may be, there is nonetheless reason to question whether they adequately describe the actual nature of the current infant mortality problem in the United States. Close examination of health data and of allied statistics suggests that the factors presently influencing America's relatively high rate of infant mortality may not only be more complex than is generally supposed but different in key respects from those in other contemporary Western societies.

Historical Perspectives

America's infant mortality problem can be placed in both international and historical perspective with the aid of data compiled by the United Nations Population Office and of estimates published by the Organization for Economic Cooperation and Development (OECD) (table 2–1). In the late 1980s, by these figures, America's infant mortality rate looked unusually high in relation to its level of per capita output

TABLE 2–1
REPORTED INFANT MORTALITY AND ESTIMATED PER CAPITA OUTPUT IN THE UNITED STATES AND FIFTEEN OTHER OECD COUNTRIES, 1925–1929 AND 1986/87

	Reported Infant Mortality Rate (per 1,000 live births)		*Estimated per Capita Output (international $, 1980 prices)*	
Country	1925–29	1986	1929	1987
United States	69.0	10.4	4,909	13,550
Australia	53.2	8.9	3,146	9,533
Austria	120.0	10.3	2,118	8,792
Belgium	101.3	16.1[a]	2,882	8,769
Canada	93.8	7.9	3,286	12,702
Denmark	82.2	8.2	2,913	9,949
Finland	89.8	5.9	1,667	9,500
France	91.4	8.0	2,629	9,475
Germany[b]	98.1	8.5	2,153	9,964
Italy	122.2	9.8[c]	2,089	9,023
Japan	140.8	5.2	1,162	9,756
Netherlands	57.9	7.8	3,373	9,197
Norway	50.4	8.0	2,184	11,653
Sweden	57.7	5.9	2,242	10,328
Switzerland	55.5	6.8	3,672	11,907
United Kingdom	73.3	9.6[d]	3,200	9,178
Unweighted average for 15 OECD countries	85.8	8.5	2,581	9,982
U.S. unweighted average	0.80	1.24	1.90	1.36

NOTES: Per capita output estimates are for gross domestic product per person, adjusted for purchasing power differences among countries and over time.
a. 1984.
b. Postwar data for Federal Republic of Germany.
c. 1985.
d. England and Wales only.
SOURCES: Angus Maddison, *The World Economy in the Twentieth Century* (Paris: Organization for Economic Cooperation and Development, 1989), p. 19; United Nations, *Demographic Yearbook 1952* (New York: UN Statistical Office, 1952), pp. 320–26; U.S. Department of Health and Human Services, *Health, United States, 1989* (Hyattsville, Md.: National Center for Health Statistics, 1990), p. 116.

(these per capita output estimates have been adjusted for purchasing power parity and may thus provide a more accurate impression of underlying economic differentials than ones standardized according to prevailing foreign exchange rates). That circumstance, however, was by no means new. Six decades earlier, in the late 1920s, the United States had also registered an unusually high rate of infant mortality in relation to its level of per capita output. If that relationship was less apparent in the earlier period, this was in part because America's estimated level of per capita output was so much higher than for any other industrialized country. Even so, a number of Western societies, including two with less than half of America's estimated contemporary level of output per capita, reported lower rates of infant mortality than the United States for the years 1925–1929.

In 1925 the welfare state was still in its infancy throughout the Western world. Sweden was the first modern state to establish such social guarantees, but the Swedish "Third Way" was barely through its first decade of experimentation by the mid-1920s. Social welfare provisions were correspondingly more modest in other Western states—including several with distinctly lower reported infant mortality rates at the time than the United States. The Netherlands offers a case in point. In the late 1920s its reported infant mortality rate was only about five-sixths the level registered in the United States. In 1929, however, total government expenditures in Holland accounted for just over 11 percent of GDP, as against 10 percent in the United States;[5] correspondingly, per capita government expenditures in Holland were no more than only three-fourths of the contemporary U.S. level.

Generally speaking, government expenditures on social security, income maintenance, and health care accounted for only a minor share of national output in Western countries before World War II.[6] Differences in public expenditures on social programs, then, would seem to be an inadequate explanation for this longstanding tendency in U.S. infant mortality trends. Whether it is to be explained by a single, continuous set of factors or by successive series of new conditions, the persistence of surprisingly high rates of infant mortality would seem characteristic of U.S. health patterns over much of the twentieth century.

Table 2–2 traces American infant mortality rates between the mid-1930s and the late 1980s and ranks American rates with those reported by the twenty-three other current members of the OECD—an organization whose membership closely approximates the current boundaries of the Western industrialized world. America's registered rate was never one of the lowest. In 1935, in its "best" showing, it ranked seventh, with a rate almost two-thirds higher than New

TABLE 2–2

REPORTED RATES FOR U.S. INFANT MORTALITY COMPARED WITH REPORTED RATES FOR OTHER CURRENT OECD COUNTRIES, 1935–1987

(rates per 1,000 live births)

	1935	*1955*	*1965*	*1970*	*1975*	*1980*	*1985*[a]	*1986*[b]	*1987*[c]
Reported U.S. rate	55.7	26.4	24.7	20.0	16.1	12.6	10.6	10.4	10.1
Lowest reported OECD rate	32.1	17.4	13.3	11.1	8.6	7.5	5.5	5.2	5.0
Country	New Zealand	Sweden	Sweden	Sweden	Japan	Japan	Japan	Japan	Japan
U.S. rank in OECD	7	8	16	15	14 (tie)	15 (tie)	19	18	—
Reported U.S. white rate	51.9	23.6	22.9	17.8	14.2	11.0	9.3	9.1	8.9
U.S. white rank in OECD	7	6	12	9	8	9 (tie)	15	14	—
Reported U.S. black rate	82.2[d]	42.8[d]	41.7	32.6	26.2	21.4	18.2	18.0	17.9
U.S. black rank in OECD	18	18	22	22	22	22	23	23	—

— = not available.

a. Ranking data for Belgium and Spain are provisional.

b. Ranking data for Spain are provisional.

c. Data taken directly from the *Japan Statistical Yearbook* on the presumption that Japan remains the OECD country with the lowest infant mortality.

d. Nonwhite.

SOURCES: For 1935: United Nations, *Demographic Yearbook 1952* (New York, UN, 1952); U.S. Bureau of the Census, *Historical Statistics of the United States: Colonial Times to 1970, Part 1* (Washington, D.C.: U.S. Department of Commerce, 1975), p. 57. For 1955–1975: *Demographic Yearbook*, various issues. For 1980–1987: U.S. Department of Health and Human Services, *Health, United States* (Hyattsville, Md.: National Center for Health Statistics): 1988 ed., p. 59, and 1989 ed., p. 116; Government of Japan Management and Coordination Agency, *Japan Statistical Yearbook 1989* (Tokyo: Statistics Bureau, 1989), p. 51.

Zealand's. Thereafter its ranking commenced a general decline until it had descended to eighteenth or nineteenth place by the mid-1980s.

As is widely known, infant mortality rates for black Americans are among the highest for any contemporary Western population. In the 1960s and 1970s only Portugal and Turkey reported higher infant mortality rates among OECD members; by the mid-1980s Portugal's was lower than the U.S. black rate.

High as U.S. black rates may be, however, they do not in themselves account for America's mediocre, and descending, standing among Western societies. Though sometimes only half the rate for black Americans, U.S. white infant mortality rates have never been among the lowest for industrialized societies. Between 1955 and 1985, moreover, the U.S. white ranking dropped sharply by comparison with other OECD countries, falling from sixth place down to fifteenth. By 1986 the U.S. white rate was reported to be higher than the national figures for such places as Ireland and Spain.

Mortality Structures

Perhaps surprisingly, America's standing among Western societies with respect to infant mortality does not correspond closely with its ranking for total mortality. Whereas the United States ranked eighteenth or nineteenth in infant mortality compared with twenty-four OECD states in the mid-1980s, its total age-standardized mortality rate in 1987, according to the World Health Organization, would have put it in eleventh place.[7] According to these estimates, total age-standardized mortality was lower in the United States than in the UK, Italy, or West Germany, all of which reported significantly lower rates of infant mortality. In a number of countries, in fact, infant mortality rates are reported to be lower than in the United States, even though life expectancy at birth is reported to be somewhat higher in the United States (table 2–3). In the mid-1980s, for example, two Scandinavian societies—Denmark and Finland—reported slightly lower levels of life expectancy at birth than did the United States. Nevertheless, the reported level of infant mortality at the time was only three-quarters the U.S. level in Denmark and less than two-thirds the U.S. level in Finland.

The paradox does not derive from unusually high rates of infant mortality in the U.S. black population. In 1986, for example, life expectancy at birth for white Americans was registered to be higher than for the national populations of such places as Finland, Denmark, the Federal Republic of Germany, Ireland, and Scotland, even though every one of these countries reported a lower rate of infant mortality.[8] By comparison with other contemporary Western societies, it would

TABLE 2–3
SELECTED COUNTRIES REPORTING LOWER MORTALITY RATES AND LOWER TOTAL LIFE EXPECTANCY AT BIRTH THAN THE UNITED STATES, CIRCA 1985

Country	*Reported Rate of Infant Mortality (per 1,000 live births)*	*Mean of Reported Male & Female Life Expectancy (years)*
Finland	6.8	74.3
Denmark	7.9	74.6
Ireland	8.8	72.9
Federal Republic of Germany	9.0	74.5
Luxembourg	9.0	73.4
Singapore	9.3	71.4
Belgium	9.4	73.4
Italy	10.3	74.4
United States	10.6	74.7

NOTES: Infant mortality data are for 1985. Life expectancy data are for the following years: Finland (1985), Denmark (1984–1985), Ireland (1980–1982), Federal Republic of Germany (1983–1985), Luxembourg (1980–1982), Singapore (1980), Belgium (1979–1982), Italy (1981), and United States (1984).
a. Arithmetic average of male and female life expectancy at birth provides a close approximation of actual total life expectancy for typical low-mortality populations but not necessarily the precise figure.
SOURCES: Derived from United Nations, *Demographic Yearbook 1986* (New York: UN, 1988), and U.S. Department of Health and Human Services, *Health, United States, 1988* (Hyattsville, Md.: National Center for Health Statistics, 1989), p. 59.

seem, American infants would appear especially vulnerable—far more so than one might expect, considering the health levels of the rest of the population. In proximate terms, American infant mortality today is defined by the peculiar, and from the standpoint of babies unfavorable, structure of American survival schedules.

Reliability of Data

How is America's high current level of infant mortality to be explained? In some measure the differential between the United States and other Western societies may be a statistical artifact. Registration of infant mortality in the United States may not yet be totally complete: in the mid-1970s, for example, some underreporting was identified in poor counties in the rural South.[9] Nevertheless, in their uniformity of definition, completeness of coverage, and consonance with recommended WHO norms, American data compare favorably with those from a number of other industrialized countries. The guidelines for WHO's International Classification of Diseases, ninth revision (ICD-9), stipulate that all births evidencing any signs of life be included for the purposes of defining infant mortality, regardless of the duration of pregnancy or the size of the newborn. In principle U.S. procedures conform to these guidelines. In 1988, for example, U.S. vital statistics registered almost 24,000 infants weighing less than 1 kilogram (about 2.2 pounds).[10] Survival rates for this high-risk cohort are extremely low. Although the group accounted for just over one-half of 1 percent of the year's registered births, it may end up accounting for nearly a third of its registered infant deaths.[11]

In Switzerland, by contrast, an infant must be at least 30 centimeters long at birth to be counted as living;[12] the restriction effectively excludes most infants weighing less than a kilogram. The country's relatively low reported rate of infant mortality reflects in part the categorical exclusion of these high-risk births. Switzerland, however, is not the only Western country to enforce its own particular definition on infant deaths. Italy, for example, has at least three definitions in force in different regions of the nation.[13] A recent study on European vital statistics in twenty-three countries concluded that "there are many indications of differences in recording and reporting live birth, fetal death, and infant death within the European region of the WHO. Even where ICD-9 recommendations are adopted as the legal definition, some countries have incomplete registration or reporting of events."[14]

In Spain, according to UNICEF's *State of the World's Children 1990*, only 1 percent of the country's children are reported to be born at low birthweight (that is, below 2,500 grams, or about 5.5 pounds).[15] By contrast, low-birthweight babies account for around 3–5 percent of the newborns in other Western European countries. Such an implausibly low proportion of high-risk births would go far in explaining why the country's current reported rate of infant mortality is the same as West Germany's and lower than those of such places as Australia and the United Kingdom.

Underregistration of infant deaths may also be indicated by the proportion of infant deaths reported for the first twenty-four hours after birth (table 2–4). In Australia, Canada, and the United States, more than a third of all infant deaths are reported to take place in the first day; in Sweden and Japan, where infant mortality rates are currently lowest, such deaths account for about a quarter of the total. In other places corresponding fractions are suspiciously low. Less than a sixth of France's infant deaths are reported to occur in the first day of life. In Hong Kong such deaths are reported to account for only

TABLE 2–4
INFANT DEATHS DURING THE FIRST DAY OF LIFE IN THE UNITED STATES AND SELECTED OTHER COUNTRIES, 1985

Country or Area	*Percent*
United States[a]	38.1
Canada	38.1
Greece	23.4
Sweden	19.8
Federal Republic of Germany	18.3
New Zealand	18.2
France	15.7
Luxembourg	10.8
Iceland	4.0
Hong Kong	4.0

a. Data for 1984; all other data are for 1985.
SOURCE: Derived from United Nations, *Demographic Yearbook 1986* (New York: UN, 1988).

one-twenty-fifth of the infant mortality total. One can credit Hong Kong with considerable achievements in public health over the past generation without subscribing to the accuracy of those specific figures.

Increased standardization of infant mortality data would undoubtedly alter America's ranking within the OECD grouping. The potential impact of such revisions, however, should not be exaggerated. America might move from the bottom third toward the middle, but it would be unlikely to advance into the top half. No plausible revisions would place the American rate under those of the five Scandinavian countries or below Holland's. Australia and Canada, whose distributions of infant death by age most closely match the United States, report rates markedly lower than those of America—lower, indeed, than for the U.S. white population. Moreover, even if Germany and France were ascribed the same fraction of infant deaths in the first day of life as the United States (an adjustment whose justification would not be universally obvious), their 1986 infant mortality rates would still have been lower. The comparatively poor health and survival prospects for American babies—white and black alike—still require explanation.

Poverty

The conventional explanation focuses upon poverty. Despite the nation's affluence, the argument runs, the United States has an unusually high proportion of children in poverty. High levels of infant mortality, in this view, are a predictable consequence. Although the argument is plausible on its face, it appears less convincing when examined.

Poverty in the United States is regularly measured by the "poverty rate," an official government index devised in the mid-1960s; see chapter 1 for more details on its genesis. (To this date no other Western government calculates such an index for its population—a fact interesting in itself.) By its construction, however, the poverty rate would seem to mismeasure material deprivation. Deprivation is characterized, and indeed defined, by inadequate consumption. The poverty rate, by contrast, is defined in relation to income—and then only to reported income in a given year. Consumer Expenditure Surveys (CES) conducted by the U.S. Bureau of Labor Statistics consistently indicate the actual consumption of households in income brackets defined to fall below the poverty line to be well above average annual stated earnings and benefits—for the lowest income brackets, typically two, three, or even four times greater.[16] The result should not be surprising: purchasing power in any given year depends not only on annual income but upon savings, drawdown of assets, loans, and a variety of other financial arrangements.

The expenditure patterns of the "poor" in America moreover raise questions about the extent to which they assess themselves to be materially needy. In 1988 less than one-fifth of the expenditures of households reporting annual incomes of $5,000 or less was allocated to food and beverages, even though all persons in this group were defined as poor.[17] The proportion of consumption allocated to food and nonalcoholic beverages in fact was lower for this American "poverty" group than for the national populations of such places as Finland, France, Ireland, Italy, and Norway in 1985.[18] All of these places, however, had infant mortality rates significantly lower than that reported for the United States in the mid- and late 1980s.

France would seem to offer a particular challenge to the notion that poverty explains America's high rates of infant mortality. In 1980 per capita output in France was estimated at about five-sixths of the American level.[19] A separate estimate for 1987 put French per capita output at only about three-fourths of the American level.[20] Figures on income distribution are notoriously problematic, but France's income distribution is consistently estimated to be one of the most unequal in Europe;[21] dispersion of incomes in France is in fact typically estimated to be greater than in the United States. Other things being equal, these two indications would lead one to expect that the proportion of the French population living below the American poverty line would be greater—perhaps considerably greater—than the corresponding proportion of Americans. Nevertheless infant mortality was reported to

TABLE 2–5
INFANT MORTALITY RATES AND ESTIMATED RATES OF CHILD POVERTY IN SELECTED INDUSTRIALIZED COUNTRIES, 1979–1981

Country	*Estimated Rate of Child Poverty (%)*	*Reported Infant Mortality Rate (deaths per 1,000 live births)*
Australia	16.9	10.0
Canada	9.6	9.6
Federal Republic of Germany	8.2	11.6
Sweden	5.1	6.9
United Kingdom	10.7	11.0
United States	17.1	11.9

NOTE: Infant mortality figures for 1981, estimated child poverty rates for various years, 1979–1981. "Poverty" is defined as the percentage of people who have adjusted disposable income below the U.S. poverty line ($5,763 for a family of three in 1979) converted into international currencies using the purchasing power parities developed by OECD. The definition of adjustable disposable income includes all forms of cash income (including the value of food stamps in the United States and of housing allowances in the UK and Sweden) and it subtracts income and payroll taxes. This definition differs slightly from the definition of income used in official U.S. calculations of poverty rates. In 1979 (the reference year for the U.S. figures) the U.S. Bureau of the Census estimated that 16.0 percent of the U.S. children were in families with incomes below the poverty line.
SOURCES: Timothy Smeeding and Barbara Boyle Torrey, "Poor Children in Rich Countries," *Science*, November 11, 1988; U.S. Department of Health and Human Services, *Health, United States, 1989* (Hyattsville, Md.: National Center for Health Statistics, 1990), p. 116.

be lower in France than in the United States throughout the 1980s—lower indeed than the reported rate for the U.S. white population.

Whatever the limitations of the American poverty line as an index of deprivation, researchers have attempted to apply it to other Western societies. The results are intriguing. According to one such effort, child poverty rates around 1980 were virtually identical in Australia and in the United States—yet the U.S. infant mortality rate at the time was almost one-fifth higher (table 2–5). For white Americans the child poverty rate would have been noticeably lower than in Australia (13.6 versus 16.9 percent), yet their infant mortality rates were also higher (11.0 versus 10.0 percent).[22] Conversely, while white Americans around 1980 would have reported a much higher rate of child poverty than was estimated for West Germany, their infant mortality rate at the time was actually reported to be slightly lower. For the countries in this small sample, differences in estimated rates of child poverty could explain only slightly more than a third of the contemporaneous differences reported in infant mortality. In view of the purported correspondence between poverty and infant mortality, that correlation would seem surprisingly weak.

Medical Care and Perinatal Survival

By the same token, the case against the availability and adequacy of American health care looks less compelling when examined in its particulars. An international comparison of perinatal mortality rates and their components would seem to attest to this (table 2–6). Perinatal mortality is defined as "fetal deaths of 28 weeks or more gestation and infant deaths of under 7 days."[23] Perinatal mortality is considered to be strongly affected by "prenatal conditions and circumstances surrounding delivery."[24] Perinatal mortality rates depend upon birthweights of fetuses or babies and upon their specific chances of surviving the perinatal period at any given weight. Thanks to the ongoing inquiries of the International Collaborative Effort on Perinatal and Infant Mortality, data are available on both birthweight distribution and perinatal mortality by weight for Japan, Norway, and the United States for the early 1980s. Table 2–6 highlights a perhaps surprising finding: perinatal mortality rates in Norway and Japan would have been dramatically lower if these populations had experienced either the U.S. or the U.S. black birthweight-specific perinatal survival rate. For Japan, perinatal mortality would have been reduced by more than a third by either of the American survival schedules.

While biological, social, and economic factors undoubtedly figure in the odds of perinatal death at any given birthweight, the quality of medical care is commonly the decisive factor—especially for the high-risk infants at low birthweights. By comparison with other Western societies enjoying especially low rates of infant mortality, U.S. babies at any given birthweight appear to have unusually good chances of surviving the perinatal period, regardless of race. All other things being equal, this would seem to suggest that medical care for infants in the United States is actually rather better than in some other advanced industrial societies.

The distribution of birthweights for babies in the United States, however, would seem to be decidedly less auspicious than in either Norway or Japan. As is widely known, the proportion of low-birthweight babies in the U.S. black population is extremely high in comparison with virtually any other Western population. With either Norway or Japan's birthweight schedules, U.S. black perinatal mortality rates in the early 1980s would have been reduced by fully two-thirds.

TABLE 2–6
ACTUAL AND ADJUSTED PERINATAL MORTALITY RATES FOR JAPAN, NORWAY, AND THE UNITED STATES, 1980–1982
(per 1,000 live births)

	Population			
			United States	
	Japan	Norway	*White*	*Black*
1. Actual reported rate	10.8	10.5	11.1	19.8
2. Own birthweights, U.S. white birthweight-specific mortality	7.9	6.8	—	19.3
3. Own birthweights, U.S. black birthweight-specific mortality	6.5	6.6	8.3	—
4. Own birthweight-specific mortality, Japanese birthweights	—	11.9	7.9	6.5
5. Own birthweight-specific mortality, Norwegian birthweights	10.7	—	6.8	6.6
Ratio				
1. = 100				
2.	73	65	—	97
3.	60	63	75	—
4.	—	113	71	33
5.	99	—	61	33

NOTE: Perinatal mortality refers to fetal deaths of twenty-eight weeks or more gestation and infant deaths under seven days. Birthweight distributions are for singleton births only. U.S. distributions and mortality rates are based on data for California, Michigan, Missouri, Upstate New York, and North Carolina.
SOURCE: Derived from Eva D. Alberman and Stephen J. W. Evans, "Trends in Birthweight Distribution: 1970–83," in U.S. Department of Health and Human Services, *Proceedings of the International Collaborative Effort on Perinatal and Infant Mortality,* vol. 2 (Hyattsville, Md.: National Center for Health Statistics, 1988), pp. III-61, III-62.

What may be less generally appreciated is the extent to which the U.S. white population suffers from a tendency toward low birthweight. With Norway's birthweight profile, for example, U.S. white perinatal mortality in the early 1980s would have been reduced by almost two-fifths. And even though the median birthweight is currently lower in Japan than in the United States, the Japanese birthweight profile

TABLE 2–7
ESTIMATED PROPORTION OF LOW-BIRTHWEIGHT BABIES, BY MATERIAL CHARACTERISTICS, 1982

Characteristic	*All Races*	*White*	*Black*
All births	6.6	5.6	12.2
Family income 149% of poverty line or less	7.7	5.6	12.9
150% or more	6.2	5.6	11.4
Child wanted at conception	5.8	5.3	10.2
Mistimed or unwanted	7.9	6.2	13.6
No smoking during pregnancy	4.2	3.3	9.0
15 or more cigarettes per day	12.9	12.6	18.6

NOTE: "Low birthweight" is defined as 2,500 grams or less.
SOURCE: E. R. Pamuk and W. D. Mosher, *Health Aspects of Pregnancy and Childbirth, United States, 1982,* series 23, no. 16 (Hyattsville, Md.: National Center for Health Statistics, 1988), pp. 52–53.

would hypothetically be associated with a reduction of white American perinatal mortality rates by almost 30 percent.

Biological and Behavioral Factors in Low Birthweight

To understand the nature of the American mortality problem, it would seem, one must account for the high incidence of high-risk, low-weight births for black and white Americans alike. Although the notion would surely evoke controversy, one cannot dismiss the possibility that the extremely high incidence of low birthweight in the U.S. black population may be caused in part by biological factors. Even after controlling for mothers' age and education,[25] and income (table 2–7), the proportion of low-birthweight babies is roughly twice as high for black as for white Americans. Recent research suggests the same may be true when one controls for the number of visits for prenatal medical care.[26] Data for the recent Caribbean and African immigrants into the United States, and for the minority populations of such places as France, Canada, and the United Kingdom, could provide additional information for an initial test of such a biological hypothesis.

Whatever proportion (if any) of low-birthweight differentials proves to be attributable to biological factors, parental attitudes and behavior apparently have a consequential influence on a child's chances of being born at a low birthweight (table 2–7). In 1982 the National

Center for Health Statistics conducted an extensive survey on pregnancy and child health in the United States. Among its objectives was to collect data on the correlates of low birthweight. According to its findings there was no measurable difference in the incidence of low birthweight for white infants born into families above or below an income line set at 150 percent of the poverty rate. (For black infants some differential was evident: babies born into the poorer families were about a seventh more likely to weigh less than 2,500 grams at birth.)

Whereas no measurable differences in the incidence of low birthweight by income were reported for white infants, white infants were about one-sixth more likely to be born into this high-risk group if their parents described them as "mistimed or unwanted at conception" rather than "wanted at conception." Among black Americans the differential was greater: babies described as "mistimed or unwanted" were fully a third more likely to be born at low birthweight than those described as "wanted at conception."

Still greater differences were associated with cigarette smoking. Among blacks, mothers who reported smoking fifteen or more cigarettes daily through pregnancy also reported an incidence of low birthweight more than twice that of black mothers who said they did not smoke during pregnancy. Among whites the differential between self-described nonsmokers and mothers who said they smoked fifteen or more cigarettes a day through pregnancy was a factor of almost four. While heavy smoking is undoubtedly not beneficial to the fetus, these differentials probably speak to more than simply the impact of tobacco. At a time when American doctors routinely and forcefully recommend against smoking during pregnancy for the good of the baby, the decision to smoke heavily during pregnancy may well be taken as a proxy for an entire locus of parental attitudes and practices that may bear upon the well-being of the infant.

Illegitimacy

Another proxy in the United States for a locus of attitudes and behaviors bearing upon the health of an infant is the decision to bear a child out of wedlock. Table 2–8 displays the incidence of low birthweight in the United States by mother's marital status. The differentials are striking. Among black Americans, the incidence of low birthweight is almost half again as high among children of unmarried mothers as for married mothers. It is fully two-thirds higher in the white American population. The incidence of low birthweight is consistently, and often considerably, higher for illegitimate babies regardless of the age or

TABLE 2–8

PROPORTION OF LOW-BIRTHWEIGHT BABIES BY MARITAL STATUS OF MOTHER, 1987

		Age of Mother						
	Total	Under 15	15–19	20–24	25–29	30–34	35–39	40–44
Total	6.9	13.7	9.3	7.1	6.1	6.2	6.9	7.9
White	5.7	10.4	7.7	5.8	5.1	5.2	6.0	7.1
Married	5.1	9.9	7.1	5.2	4.7	4.9	5.6	6.5
Unmarried	8.5	10.5	12.3	8.0	8.8	9.4	10.3	11.1
Black	12.7	16.2	13.1	12.3	12.5	13.0	13.4	12.9
Married	9.9	10.9	10.9	9.6	9.6	10.2	11.1	10.9
Unmarried	14.4	16.2	13.3	13.6	15.8	17.3	17.6	16.4

NOTE: "Low birthweight" is defined as 2,500 grams or less.
SOURCE: U.S. Department of Health and Human Services, *Vital Statistics on the United States: 1987* (Hyattsville, Md.: National Center for Health Statistics, 1989), p. 255.

race of the mother. So substantial are the differentials associated with illegitimacy that an American baby born to a teenage mother is less likely to register low birthweight if the mother is married and black than if she is unmarried but white.

American specialists on public health commonly hold that the association between illegitimacy and higher rates of infant mortality is spurious. A recent editorial note in the *Morbidity and Mortality Weekly Report* (MMWR) of the Centers for Disease Control, for example, stated that "the marital status of the mother confers neither risk nor protection to the infant; rather, the principal benefits of marriage to infant survival are economic and social support."[27] In the recent past, when exceedingly few data were available on the social correlates of illegitimacy in America, such a viewpoint was perhaps intuitively plausible. In light of newly gathered data, however, it no longer looks tenable.

Since the early 1980s the NCHS has been conducting a project linking births and infant deaths for the entire country and tabulating data on infant deaths against various maternal characteristics. An eight-state pilot survey in 1982 cross-tabulated infant mortality rates not only against the race and age of the mother but also against her marital status and level of education (table 2–9). According to these data infant mortality rates for white mothers over age twenty were higher for unmarried but college-educated women than for married

TABLE 2–9
INFANT MORTALITY RATES BY RACE, EDUCATION, AND MARITAL STATUS FOR WOMEN AGED TWENTY OR OLDER IN EIGHT PILOT STATES, 1982 U.S. BIRTH COHORT
(per 1,000 live births)

Years of Education	*All Races*	*White*	*Black*
All women	10.5	9.0	21.9
0 to 8	13.0	12.0	27.5
9 to 11	16.7	12.9	29.1
12	10.1	8.8	20.0
13 to 15	9.5	8.1	18.6
16+	8.6	7.9	20.6
Married women	9.1	8.4	18.2
0 to 8	11.1	10.6	24.5
9 to 11	12.4	11.1	24.9
12	8.9	8.4	16.4
13 to 15	8.5	7.8	16.8
16+	8.3	7.9	19.2
Unmarried women	20.7	16.2	25.2
0 to 8	20.7	18.1	29.8
9 to 11	24.9	19.2	30.7
12	18.6	14.6	23.1
13 to 15	18.9	15.1	20.9
16+	18.4	11.6	26.8

NOTE: The eight pilot states are Illinois, Indiana, Massachusetts, Michigan, Missouri, New Hampshire, Vermont, and Wisconsin. "All races" includes races other than black and white.
SOURCE: National Center for Health Statistics, Pilot Linked Birth and Infant Death File, unpublished data.

high school—or even grade school—dropouts. The same pattern held true among black mothers.

It is not obvious that such disparities should be attributed to poverty. In the past, Census Bureau reports on poverty have not disaggregated poverty rates for various types of families by the educational level of the household head. For the first time, however, the advance report on poverty in America for 1988 did so. According to its findings, the 1988 white, family poverty rate for households with children headed by a woman with a year or more of college education was 18.1 percent. By contrast the corresponding poverty rate for married-couple families headed by a person with less than a high school education was estimated to be 19.0 percent.[28] For black Americans the findings were similar: corresponding family poverty rates were reported to be 24.3 percent and 28.0 percent, respectively.[29] Although these categories do not correspond precisely with the categories outlined in table 2–9, they are close enough to suggest that poverty, as officially defined, may not explain the high rates of infant mortality for educated but unmarried mothers.

Parental Practices and Infant Health

Such findings, one may note, are broadly consistent with other recent data that have been compiled on the health of American children. In analyzing the results of the NCHS 1988 National Health Interview Survey on Child Health, Deborah Dawson found that "health vulnerability . . . scores were elevated" for children in single-parent homes, even after controlling for race, income, and maternal education.[30]

Why in particular should the health of infants and children in single-parent homes be worse than race, age, education, or income would seem to predict? A factor that may go far in explaining such differentials is the attitude of the single parent toward the care and treatment of offspring. For the population as a whole the number of maternal visits for prenatal care during pregnancy may arguably be used as a serviceable, if obviously imperfect, indicator of such attitudes. Unpublished data from the NCHS Linked Birth and Infant Death Files show a strong correlation between infant mortality rates and the number of prenatal medical visits by the pregnant mother (table 2–10).

In 1983 infant mortality for black mothers who reported no prenatal visits for medical care was reported to be more than three times the U.S. black national average. Among white Americans the infant mortality rate for such mothers was almost four times the corresponding national average. For both black and white infants, infant mortality rates were less than one-seventh as great among mothers reporting thirteen to sixteen prenatal visits as for those who sought or received no prenatal care. Although total infant mortality rates were almost twice as high for U.S. blacks as U.S. whites in 1983, infant mortality rates for black mothers who reported sixteen prenatal medical visits were actually *lower* than the white national average.

Education and income levels are likely to correlate positively with the number of prenatal medical visits a mother makes. Maternal age may also be a factor. And high-risk, premature births will likely correspond with low numbers of prenatal visits by virtue of their short gestation. Nevertheless the relationship between infant mortality and the frequency of prenatal medical visits is so robust that a strong

TABLE 2–10
INFANT MORTALITY RATES BY RACE OF MOTHER AND NUMBER OF PRENATAL VISITS IN THE UNITED STATES, 1983 BIRTH COHORT
(deaths per 1,000 live births)

Number of Prenatal Visits	*White*	*Black*
All women	9.4	19.0
None	35.5	58.9
1–4	43.2	52.4
5–8	16.8	20.4
9–12	5.6	9.0
13–16	4.6	7.8
17 or more	23.1	36.1

SOURCE: National Center for Health Statistics, Pilot Linked Birth and Infant Death File, unpublished data.

correspondence may be expected to remain even after controlling for such exogenous influences.

While prenatal medical care is surely more important in itself, a parent's interest in it is also suggestive of a broader set of outlooks and practices. Unmarried mothers are decidedly less likely to seek medical care for the infants they are bearing. Among black Americans, illegitimate children in 1987 were nearly four times as likely to have received no prenatal care whatsoever as babies born into marital unions. For white infants in the same year, illegitimate babies were about five times as likely to have received no prenatal care as ones born in wedlock.[31] Although further research is required to answer the question conclusively, social and economic factors alone seem highly unlikely to account for such disparities.

In recent years, illegitimacy has played a role in retarding improvements in medical care for unborn children. According to a recent supplement to the NCHS *Monthly Vital Statistics Report (MVSR)*, "much of the lack of improvement in early receipt of prenatal care (between 1979 and 1988) is associated with the increasing proportion of births to unmarried mothers who are less likely than married mothers to begin care early."[32]

Since 1960 childbearing in the United States has been profoundly affected by a sweeping change in patterns of illegitimacy. Although fewer children were born in 1988 than in 1960, the number of births reported for unmarried mothers rose at an average pace of more than 5 percent a year over this period. By 1988 more than a quarter of all births in America were to unmarried women. As the *MVSR* recently noted, the profile of the unmarried mother has changed with the increase in illegitimate births:

> [In 1988] increases were greater for white women, continuing a pattern that has been observed in recent years. . . . Rates of nonmarital childbearing and proportions of nonmarital births among black women continue to be substantially higher than among white women. In recent years, however, because the increases have been much more rapid for white than black women, the differentials by race have diminished. . . . In recent years . . . the largest increases in nonmarital childbearing have occurred among relatively older women. One third of unmarried mothers were 25 years and older in 1988 as compared with 24 percent in 1980. . . . The pattern of more rapid increases in rates for older than for younger unmarried mothers is generally replicated in rates for white and black women, but the pace of increase has been greater for white women.[33]

By 1988, stereotypes notwithstanding, black teenage mothers accounted for less than one-seventh of the illegitimate births in the United States. Nearly a quarter of the children born to white women in their early twenties were out of wedlock. And although illegitimacy ratios varied considerably by state, there was no state in the Union in which unmarried mothers accounted for as few as a tenth of the total white births.[34]

U.S. Illegitimacy in International Perspective

Compared with other Western societies, illegitimacy ratios in the United States today would not appear to be exceptional. Illegitimacy has been reported on the rise in virtually all Western countries over the past generation. While the reported ratios in the United States in the mid- to late 1980s were more than twice as high as in West Germany and more than four times as high as in Italy, they were roughly the same as in France and the United Kingdom and less than half that in Sweden, the society with the lowest reported infant mortality rate in Europe. Different patterns of parental behavior, however, appear to be associated with illegitimacy in Sweden and in the United States. In Sweden, despite a reported illegitimacy ratio of almost 50 percent, only 17 percent of family households in 1988 were single-parent households; consensual union or cohabitation appears to have been common. In the United States, by contrast, about 23 percent of all families with children were single-parent households in 1988 (table 2–11). The United States in fact appears to have much the highest proportion of single-parent families of any major Western country. Comparative data on

TABLE 2–11
PERCENTAGE OF SINGLE-PARENT FAMILIES AMONG FAMILY HOUSEHOLDS WITH CHILDREN IN SELECTED COUNTRIES, 1982–1988

Country	*Percent*
Australia (1982)	11.4
Canada (1986)	14.8
Federal Republic of Germany (1988)	13.5
France (1988)	10.9
Japan (1985)	5.9
Sweden (1985)	16.9
United Kingdom (1987)	12.7
United States (1988)	22.9

NOTE: "Children" is defined as under eighteen years of age with the following exceptions: Australia includes all children and full-time students aged fifteen–twenty. The UK includes all children under sixteen and full-time students sixteen–seventeen years; data exclude Northern Ireland and are based on a household survey. Definitions of households, children, and treatment of cohabitation may differ across countries.
SOURCE: Frank Hobbs and Laura Lippmann, "Children's Wellbeing: An International Comparison" (U.S. Census Bureau Center for International Research, February 1990), table 7.

the actual proportion of children living in single-parent families are more difficult to come by. Initial inquiries, however, suggest that that proportion is much higher in the United States today than in a number of countries with similar or higher reported illegitimacy ratios (table 2–12). An unusually high proportion of children in single-parent families moreover appears to have been characteristic of the United States since at least the 1960s.

The nature of the American infant mortality problem can be elucidated by comparison between the circumstances of the population of Norway, a virtually homogenous ethnic European population, and that of the white population of the United States. Throughout the 1980s, reported infant mortality rates have been lower for Norwegians than for white Americans—often substantially lower. This is not because perinatal survival rates by birthweight—a quantity decisively affected by medical care—were higher in Norway. To the contrary: if white American infants in the early 1980s had been exposed to Norwegian survival schedules, their reported rates of perinatal mortality would have been even higher. Nor would white Americans have benefited from Norwegian mortality rates during the postneonatal period (months two–eleven—a rate that is commonly believed to be especially sensitive

TABLE 2–12
PERCENTAGE OF CHILDREN IN SINGLE-PARENT FAMILIES IN SELECTED INDUSTRIALIZED COUNTRIES, 1960–1986

Country	*1960*	*1970*	*1975*	*1980*	*1983–86*
Canada	NA	NA	10.5	12.8	NA
Norway	NA	NA	8.1	10.9	13.9
Sweden	7.8	12.0	11.4	13.5	NA
United Kingdom	NA	8.0	10.0	12.0	14.0
United States	9.1	11.9	NA	19.7	23.4

NOTES: All data for United Kingdom refer to Great Britain. The 1983–1986 data for the UK refer to 1986, to 1983 for Norway, and to 1985 for the United States. Children are defined as follows: Canada—age zero to twenty-four years; Norway—under age twenty; Sweden—eighteen years and under for 1960, 1970, and 1975 and fifteen years and under for 1980; United Kingdom—under age sixteen or aged sixteen–eighteen and in full-time education; United States—under eighteen. NA = not available.
SOURCE: Christopher Jencks and Barbara Boyle Torrey, "Beyond Income and Poverty: Trends in Social Welfare among Children and the Elderly since 1960," in John L. Palmer, Timothy Smeeding, and Barbara Boyle Torrey, eds., *The Vulnerable* (Washington, D.C.: Urban Institute, 1988), p. 257.

to a population's socioeconomic conditions). In 1986 Norway's postneonatal mortality rate was reported at 3.4 per thousand, as against 3.1 for American whites. If Norway had been an American state, it would have ranked thirty-fifth in white postneonatal mortality: just above Mississippi but below West Virginia.[35] Norway's comparatively high rates of postneonatal mortality are perfectly consistent with economic data pertaining to the country's living standards. In 1980, for example, per capita consumption in Norway was placed at just over two-thirds of the American level in one comprehensive study.[36] In relation to average consumption levels of the American white population alone, it would have been even lower.

It would appear, in short, that white America's higher rates of infant mortality are explained not by poverty (as conventionally construed) or by medical care but rather by the habits, actions, and indeed life styles of a critical portion of its parents. The life style or behavior patterns attendant upon the American mode of illegitimacy appear to be particularly injurious for newborn infants. The distinctive etiology of illegitimacy in the United States may be appreciated when one remembers that reported ratios for illegitimacy in the 1980s were only about half as high for white Americans as in Norway. If illegitimacy ratios had been twice as high for white Americans in the 1980s, there can be little doubt that their rates of infant mortality would have been even higher.

Data at the Local Level

The contribution of parental behavior or life styles to American levels of infant mortality is further emphasized in examination of data at the local level. Considerable variations both in socioeconomic characteristics and patterns of family formation are reported between the fifty states and the District of Columbia. These variations tend to highlight patterns obtaining for the nation as a whole.

In the mid-1980s the unweighted average rate of infant mortality for the five areas reporting the highest levels was more than 60 percent higher than for the five lowest (table 2–13). Per capita personal income, however, was officially estimated to be *greater* in the areas with the higher reported rates of infant mortality than in the states with the lowest reported rates. Indeed, per capita personal income in the lowest infant mortality states was officially estimated to be below the national average, and above the national average in the highest infant mortality areas. Per capita personal income data are not adjusted to reflect differences in price levels or in the cost of living; in fact, the U.S. government did not even prepare estimates of those particular differences during the 1980s. Purchasing power adjustments would doubtless affect the results in table 2–13 to some degree. Even so, differences in per capita income do not appear to explain much of the differential in local American infant mortality rates.

Conversely, even without adjustments, differences in illegitimacy ratios correspond with differences in infant mortality at the state level (table 2–14). The unweighted average for the five highest infant mortality areas in the mid-1980s was more than twice as high as for the five lowest mortality states.

Data for 1980 are somewhat more comprehensive insofar as the decennial census was conducted in that year (table 2–15). Characteristics conventionally thought to explain health differential offer little insight into the gap between high and low infant mortality states. For 1979–1981 the unweighted average for the five highest infant mortality areas exceeded that of the five lowest infant mortality states by more than three-fourths (77 percent). Yet officially estimated per capita income was below the national average in the low infant mortality states and above the national average in the high infant mortality areas. The estimated family poverty rate was higher for the high infant mortality group—but it was also above the national average for the states with the lowest infant mortality rates. Similarly unemployment rates were

TABLE 2–13

ESTIMATED AVERAGE PER CAPITA PERSONAL INCOME IN STATES AND DISTRICT WITH HIGHEST AND LOWEST REPORTED RATES OF U.S. INFANT MORTALITY, 1984–1986

State or District	*Reported Infant Mortality Rate, 1984–86 (per 1,000 live births)*	*Per Capita Personal Income, 1985 (U.S. $, 1985)*
Lowest		
North Dakota	8.4	11,961
Maine	8.8	11,834
Massachusetts	8.8	16,324
Iowa	8.9	12,554
Minnesota	9.0	14,142
(Unweighted average)	(8.8)	(13,263)
Highest		
District of Columbia	21.0	17,974
Alabama	12.9	10,752
Georgia	12.7	12,642
Delaware	12.4	14,544
Illinois	12.0	14,734
(Unweighted average)	(14.2)	(14,129)
U.S. average	10.6	13,908

NOTE: "Personal income" is "the income received by persons from all sources, that is from participation in production, from both government and business transfer payments, and from government interest" (*Local Area Personal Income*, p. xxix).

SOURCES: U.S. Department of Health and Human Services, *Health, United States, 1988* (Hyattsville, Md.: National Center for Health Statistics, 1989), p. 55; U.S. Department of Commerce, *Local Area Personal Income 1982–87*, vol. 1 (Washington, D.C.: Bureau of Economic Analysis, 1989), p. 226.

on average slightly lower for the low infant mortality states; they were no higher than the national average for the group with the highest rates of infant mortality. One may make the argument that health differentials are by definition proof of inadequate medical care expenditures in the regions with higher mortality. The fact of the matter, however, is that per capita health care expenditures were already more than 40 percent higher for the high infant mortality areas than for the low infant mortality states.

Of the economic characteristics listed in table 2–15, only the

TABLE 2–14
ILLEGITIMACY RATIOS AND PROPORTION OF CHILDREN RECEIVING AFDC PAYMENTS IN STATES AND DISTRICT WITH HIGHEST AND LOWEST REPORTED RATES OF U.S. INFANT MORTALITY, CIRCA 1985

State or District	*1984–86 Reported Infant Mortality Rate (per 1,000 live births)*	*1985 Ratio of Illegitimacy (per 1,000 live births)*	*1985 Proportion of Children Receiving AFDC Payments (%)*
Lowest			
North Dakota	8.4	115	4.3
Maine	8.8	178	11.1
Massachusetts	8.8	184	11.1
Iowa	8.9	136	10.0
Minnesota	9.0	151	8.5
(Unweighted average)	(8.8)	(153)	(9.2)
Highest			
District of Columbia	21.0	567	32.9
Alabama	12.9	249	9.4
Georgia	12.7	257	9.9
Delaware	12.4	262	10.0
Illinois	12.0	257	15.9
(Unweighted average)	(14.2)	(318)	(15.6)
U.S. average	10.6	220	11.4

NOTE: Figures for Aid to Families with Dependent Children are monthly averages rather than annual figures, and may therefore somewhat overstate annual program participation.
SOURCES: Derived from U.S. Department of Health and Human Services, *Health, United States, 1988* (Hyattsville, Md.: National Center for Health Statistics, 1989), pp. 48, 55; idem., *Social Security Bulletin*, Annual Statistical Supplement 1987, p. 295; U.S. Department of Commerce, *Statistical Abstract of the United States: 1987* (Washington, D.C.: Bureau of the Census, 1987), p. 23; *1988*, p. 63.

proportion of children receiving government benefits through the Aid to Families with Dependent Children (AFDC) program corresponds with infant mortality differences in a significant fashion. The AFDC program is itself significant insofar as it has become a vehicle for financing illegitimacy and maintaining the families of unmarried mothers. By 1987 more than half the children receiving AFDC benefits qualified for the program because their parents were unmarried; their total numbers accounted for more than nine-tenths of all children for

TABLE 2–15: ECONOMIC CHARACTERISTICS OF STATES AND DISTRICT WITH LOWEST AND HIGHEST REPORTED RATES OF U.S. INFANT MORTALITY, 1979–1981

State or District	*Reported Infant Mortality Rate, 1979–81 (per 1,000 live births)*	*Unemployment Rate, 1980*	*AFDC, 1980*	*Per Capita Personal Income, 1980 (U.S. $)*	*Per Capita Health Expenditures, 1980 (U.S. $)*	*Family Poverty Rate, 1979 (% of all families)*
Lowest						
Vermont	9.0	6.4	12.6	8,578	778	8.9
Maine	9.9	7.7	15.6	8,224	870	9.8
New Hampshire	10.0	4.7	7.3	9,789	759	6.1
Idaho	10.0	7.9	5.5	8,570	695	9.6
Hawaii	10.0	5.0	17.8	10,616	935	7.8
(Unweighted average)	(9.3)	(6.3)	(11.8)	(9,155)	(807)	(8.4)
Highest						
District of Columbia	24.1	7.2	51.2	12,279	2,198	15.1
Delaware	14.8	7.7	17.1	10,241	912	8.9
Illinois	14.6	8.3	18.2	10,840	1,033	8.4
Louisiana	14.5	6.7	14.7	8,682	857	14.4
Georgia	14.4	6.4	12.3	8,350	843	13.2
(Unweighted average)	(16.5)	(7.3)	(22.7)	(10,078)	(1,169)	(12.0)
U.S. average	12.6	7.1	14.5	9,919	958	7.4

SOURCES: U.S. Department of Health and Human Services, *Health, United States 1988* (Hyattsville, Md.: National Center for Health Statistics, 1989), pp. 55, 191–92; idem, *Social Security Bulletin, Annual Statistical Supplement 1981*, p. 250;U.S. Department of Commerce, *1980 Census of the Population* (Washington, D.C.: Bureau of the Census, 1982), various volumes; idem, *Statistical Abstract of the United States: 1981* (Washington, D.C.: Bureau of the Census, 1981), p. 380; *Statistical Abstract of the United States: 1988* (Washington, D.C.: Bureau of the Census, 1988), p. 417; idem, *County and City Data Book: 1983* (Washington, D.C.: Bureau of the Census, 1983), pp. 2–3.

American never-married mothers.[37]

Quite possibly race plays a confounding role in these comparisons at the state level insofar as black American infant mortality rates are so much higher than white rates and black Americans constitute a widely varying proportion of the total population of the different states. Census data for 1980 allow us to compare high and low infant mortality areas within the American black population by itself (table 2–16). For the five states reporting the lowest rates of black infant mortality, the unweighted average was less than three-fifths the level of the high mortality group. Median family income was virtually identical for the high and low infant mortality groups and above the black national average for both. While the family poverty rate was notably lower for the low infant mortality group, it was still distinctly below the black national average in the high infant mortality group. Median years of education of the household head was above the black national average for both groups. A distinct difference between the two groups, however, was in illegitimacy ratios: in 1980 the proportion of children born out of wedlock in the highest infant mortality areas was more than 50 percent higher than in the lowest infant mortality states. The illegitimacy ratio in the highest infant mortality areas moreover was significantly higher than the black national average.

Policy Interventions and Their Limits: Income Support, Medical Care

If parental life styles and family formation patterns play a direct and important role in determining infant survival chances, the prospects for reducing American infant mortality rates through public income support and health care policies may be less substantial than is sometimes supposed. Given the complex interactions between illegitimacy and infant mortality, estimating the extent—if any—to which expanding AFDC benefits might reduce infant mortality rates would be an intricate calculation indeed.

Parallel questions arise about the expansion of public health care programs. It is widely argued today that poorer Americans have inadequate access to health care. In much of this discussion, however, a critical distinction is obscured: the difference between availability and utilization. As long as health care treatment is a voluntary option for the potential patient—or the potential patient's parents—utilization will depend not only upon availability but also upon the attitudes, inclinations, and preferences of individual decision makers.

These attitudes and preferences bring an interesting perspective

TABLE 2–16

CHARACTERISTICS OF CHILDREN, FAMILIES, AND PERSONS IN STATES AND DISTRICT WITH HIGHEST AND LOWEST REPORTED U.S. INFANT MORTALITY RATES AMONG BLACKS, CIRCA 1980

	1979–80 Infant Mortality (per 1,000 live births)	*1979 Median Family Income ($ 1979)*	*1979 Family Poverty Rate (%)*	*1980 Education (25 years+)*	*1980 Illegitimacy Ratio (per 1,000)*
Lowest					
Hawaii	11.4	12,764	11.5	12.9	11.4
Washington	14.9	15,833	17.3	12.7	40.7
Colorado	15.4	15,732	18.7	12.7	40.3
Massachusetts	16.4	13,249	23.8	12.4	52.9
California	17.1	14,887	26.7	12.6	52.8
(Unweighted average)	(15.0)	(14,493)	(16.4)	(12.7)	(39.6)
Highest					
Delaware	27.4	13,127	25.1	12.0	66.1
District of Columbia	26.3	16,362	18.6	12.2	64.3
Illinois	25.9	14,478	27.0	12.2	66.0
Michigan	24.0	15,817	23.5	12.2	52.7
West Virginia	23.2	12,318	21.4	12.1	52.0
(Unweighted average)	(25.4)	(14,420)	(23.1)	(12.1)	(60.2)
U.S. average	21.0	12,598	26.5	12.0	55.3

SOURCES: U.S. Department of Health and Human Services, *Vital Statistics of the United States*, vol. 1 (Hyattsville, Md.: National Center for Health Statistics, 1983), p. 37; U.S. Bureau of the Census, *State and Metropolitan Area Data Book: 1986* (1986), p. 509; U.S. Department of Health and Human Services, *Health, United States:1988* (Hyattsville, Md.: National Center for Health Statistics, 1989), p. 55; U.S. Department of Commerce, *1980 Census of Population* (Washington, D.C.: Bureau of the Census, 1981), various volumes.

to the current debate about affordability of health care in contemporary America. In 1985–1987, according to a recent NCHS study, white children averaged 4.5 physician contacts a year—more than 60 percent more than for black children (table 2–17). The difference is not totally explained by poverty or income level. Black families were on average poorer than white families, and more affluent families tended to have more physician contact for their children. But the average number of

TABLE 2–17

AVERAGE ANNUAL NUMBER OF PHYSICIAN CONTACTS FOR U.S. CHILDREN UNDER EIGHTEEN YEARS OF AGE BY SELECTED CHARACTERISTICS, 1985–1987

	Race of Child	
	White	Black
All children	4.5	2.8
Family income		
Under $10,000	4.4	3.2
$10,000 to $19,999	4.1	2.5
$20,000 to $34,999	4.7	2.8
$35,000 or more	5.0	3.3
Poverty status		
In poverty	3.9	3.1
Not in poverty	4.7	2.8
Children assessed in fair or poor health	15.5	6.5
Family income		
Under $10,000	13.6	6.2
$10,000 to $19,999	12.8	6.8
$20,000 to $34,999	19.6	11.3
$35,000 or more	21.0	7.1
Poverty status		
In poverty	13.0	7.0
Not in poverty	17.5	7.0

SOURCE: P. Ries, *Health of Black and White Americans, 1985–87*, series 10, no. 171 (Hyattsville, Md.: National Center for Health Statistics, 1990), pp. 58, 62.

physician contacts was higher for whites at every income level. For families with incomes of $35,000 or more, white children averaged 5.0 contacts a year—half again as many as black children in the same income grouping. Distinctions were even more dramatic for children whose health parents described as "fair or poor." In this group white children averaged two and a half times as many average physician contacts as black children; in families with incomes above $35,000, white children averaged nearly three times as many contacts with physicians as their black counterparts. It is well known that the net worth of black households tends to be lower than that of white households in the contemporary United States, and wealth may play

TABLE 2–18
PATTERNS OF CONSUMER EXPENDITURE BY REPORTED HOUSEHOLD INCOME AND AGE OF HEAD OF HOUSEHOLD, 1988
(U.S. $, 1988)

	Reported Household Income			
Age of Household Head	Less than $5,000	$5,000–9,999	$10,000–14,999	$15,000–19,999
Under 25 years				
Health expenditures	177	207	484	625
Entertainment	450	649	896	1,053
Alcohol and tobacco	282	465	529	614
25 to 34 years				
Health expenditures	247	349	471	613
Entertainment	510	496	796	1,007
Alcohol and tobacco	359	461	472	574
35 to 44 years				
Health expenditures	749	448	808	957
Entertainment	595	564	786	1,002
Alcohol and tobacco	419	454	403	512
Memorandum Items				
Average number of children under 18 per houshold				
Under 25 years	0.2	0.5	0.4	0.4
25 to 34 years	0.9	1.4	1.1	1.1
35 to 44 years	1.1	1.5	1.4	1.3
Ratio of expenditures on entertainment, alcohol, and tobacco to health care (Health care = 100)				
Under 25 years	414	538	294	267
25 to 34 years	352	274	269	258
35 to 44 years	135	227	147	158

SOURCE: U.S. Department of Labor, Bureau of Labor Statistics, Consumer Expenditure Survey, unpublished tables.

a role in health care decisions. Nevertheless, at any given income level, white U.S. parents seemed to find health care to be more "affordable" for their children than black U.S. parents.

The Bureau of Labor Statistics Consumer Expenditure Survey reveals still more about popular perceptions of the affordability of current health care services. Table 2–18 details reported expenditures in 1988 on health care and on entertainment, alcohol, and tobacco for households reporting incomes of less than $20,000 by age of household head. Generally speaking, households in this lower half of the income scale indicated that they could "afford" to spend several times as much on entertainment, alcohol, and tobacco as on health care. It is hard to know exactly how much confidence to place in these reported expenditures; for one thing, many respondents have received free medical care under the Medicaid program. But it would appear that among these households the income elasticity of expenditure was substantially higher for household health care than for entertainment, alcohol, and tobacco. To judge by the preferences revealed in this survey, there appeared to be a tendency on the part of lower-income consumers to treat health care as an optional but dispensable luxury good whereas entertainment, alcohol, and tobacco seemed to be treated more like necessities, "affordable" at any level of household income.

As long as health care treatment is the prerogative of the patient—or the patient's parents—some fraction of the population will always deem treatment to be "unaffordable," regardless of its actual nominal cost. In such circumstances, the only way to ensure adequate coverage would be to deny the patient an option for treatment—for example, by making routine assessments mandatory, independent of the preference of the patient, and proscribing nonnegotiable therapies on their basis. One may doubt that such a proposal would be favorably received by the U.S. public at large.

Concluding Observations

More than citizens in many other societies, Americans pay close attention to their individual freedoms and rights. For better or worse one of those rights—within broad limits—is the right to be a negligent parent. One may feel that too many parents and parents-to-be misuse this particular freedom. Yet simple remedies to the injuries wrought by such abuses are not immediately evident. Even when American parents are demonstrably abusive to their chidren, the state's options are hardly ideal; can the state proceed more confidently against parents who merely expose their fetuses or infants to *potential* risks without malice aforethought? Still less obvious is the role that public policy should play in this problem if it aims to avoid creating new injuries in the course of redressing existing ones.

3

Health, Nutrition, and Literacy under Communism

From the October Revolution in 1917, until the collapse of the Soviet empire in 1989–1991, intellectuals and idealists, both in the West and in what was once called underdeveloped areas, were engaged by the claims made on behalf of the Marxist-Leninist style of governance. Not surprisingly, the promises offered by this system have intrigued and excited persons who favored purposeful action to create a better world. Marxist-Leninists argued that they, and they alone, possessed a scientific insight into the causes of poverty and injustice in the modern world. Given command of the machinery of state, they insisted, they would successfully engineer a radical transformation in the workings of the economy and of society. Through purposeful and planned collective action, they said, material advance would be dramatically accelerated, exploitation would be eliminated, and want would become a thing of the past.

In the seventy years since these promises were first officially proclaimed, governments describing themselves as Marxist-Leninist came to power over a growing fraction of the human population: by the mid-1980s, perhaps a third of mankind lived under regimes that called themselves Communist. (Although differences could be discerned among both their foreign and domestic policies, it is nevertheless appropriate to discuss them collectively, for they were united by adherence to a single and distinct theory of state power.) With the establishment of Communist rule over the Soviet Union, Mongolia, Eastern Europe, North Korea, China, Vietnam, and Cuba, it is no longer necessary to talk about Marxist-Leninist government in hypothetical terms. Each of the aforementioned experiments has had a run of more than a quarter of a century. It is not premature to judge them by their own results.

As social and economic results have come in, the claims made on behalf of Marxist-Leninist rule have in all but the most doctrinaire quarters progressively diminished. It is no longer seriously suggested that the Communist constitutional arrangement affords political freedoms or legal guarantees unavailable to populations in bourgeois societies. Nor is it widely argued in this day and age that the socialist mode of economic organization has any special efficacy in hastening material advances, spurring improvements in productivity, or promoting technological innovation.

Admirers and friendly observers have thus come to make a more limited, and seemingly careful, case for the Marxist-Leninist state. While typically conceding that the Marxist-Leninist state fares poorly in the realm of political liberty, and perhaps acknowledging that its superiority in the realm of production is less than self-evident, the argument, as currently constituted, focuses attention upon progress in combating poverty. In this realm, it is stated, Communist governments have registered exemplary accomplishments. These accomplishments, it is said, have been particularly striking with respect to the manifestation of material deprivation, especially ill health, undernutrition, and illiteracy. Through extensive and all-embracing state medical service, radical land reform, comprehensive rationing, mass education and anti-illiteracy campaigns, and other characteristic practices or policies, Communist governments are said to have done more to alleviate mass poverty than non-Communist regimes—even when the economic base at their disposal was smaller or weaker.

This argument has made a deep impression upon many intellectuals and leaders who consider themselves persons of conscience. The notion that Marxist-Leninist states are more successful than others in dealing with material poverty has been treated in many circles as received wisdom (even among people who think of themselves as anti-Communist). But strangely there seems to be little corresponding interest in the facts of the case. The specific circumstances of public health, popular nutrition, and mass education in different Communist lands are often ignored, even when relevant facts are readily available.

Surely it is more informative to examine the actual record of Marxist-Leninist states in dealing with material poverty than to rely upon generalization or assertions. This chapter reviews the performance of Communist states in three areas where they have claimed excellent results: health, nutrition, and education.

Health

More than most social indicators, health statistics provide a meaningful measure of a population's material well-being. Better health comes close to being a universally desired attribute. Health levels are understood to reflect upon living conditions in general. Changes in health, moreover, are less ambiguous than changes in economic activity: it is considerably more difficult to ascribe value to a changing mix of production than to determine whether a person is alive or dead.

Marxist-Leninist states have long affirmed their commitment to the cause of improving public health. Well before the notion of "human capital" became familiar to economists in the West, Josef Stalin had declared that "human beings are the most important and decisive capital in the world."[1] In consonance with such pronouncements, the Soviet state devised and erected an enormous public health apparatus, employing vast numbers of doctors and medical personnel who were to have extensive contact with the population at large and to provide their services at no direct cost to their patients.

The Soviet Model. The Soviet health system is the model after which medical systems in all other Communist countries, to a greater or lesser degree, have been fashioned or restructured. It was the Communist public health system that had been longest in operation. It is therefore particularly significant that Soviet health appears to have undergone unambiguous, and sustained, *deterioration* during the past several decades.

The deterioration in Soviet health conditions is represented most vividly by death rates. During the 1950s, mortality rates in the USSR registered rapid declines. According to Soviet statistics, however, these declines came to a halt in the early and mid-1960s. Thereafter death rates began to rise. The first group affected was middle-aged men. But within a matter of years death rates were rising for virtually every adult age group (see table 3–1). By the early 1970s recorded rates of infant mortality were also rising: according to official figures Soviet infant mortality jumped by about 25 percent between 1971 and 1974.[2]

What happened in the years after 1974 is to a considerable degree a matter of speculation. Starting in the mid-1970s, Soviet authorities began to impose a blackout over data pertaining to the country's health. During the darkest days of this blackout (the early and mid-1980s), it was difficult to obtain even such pedestrian numbers as the country's crude death rate. (As it happens, that figure rose from 6.9 per 1,000 in 1964 to 10.8 per 1,000 in 1984: a 56 percent jump, far too great a rise to be explained by the aging of the general population.)[3]

Only recently have the dimensions of the deterioration in health

TABLE 3–1
REPORTED INCREASES IN MORTALITY BY AGE GROUP IN THE SOVIET UNION, 1961–1976
(death rate per 1,000, both sexes)

Age Group	*Low Year*	*Death Rate*	*High Year*	*Death Rate*	*% of Increase*
20–24	1967	1.5	1976	1.7	11
25–29	1964	2.0	1971	2.2	10
30–34	1964	2.5	1976	3.0	20
35–39	1961	3.0	1976	3.8	27
40–44	1961	3.7	1976	5.3	43
45–49	1965	5.0	1976	6.9	38
50–54	1961	7.5	1976	9.3	24
55–59	1964	10.7	1976	13.4	25
60–64	1961	16.7	1976	18.9	13
65–69	1964	24.1	1976	28.0	16
70+	1961	63.0	1973	75.5	20

SOURCE: Nick Eberstadt, *The Poverty of Communism* (New Brunswick, N.J.: Transaction Publishers, 1988), p. 39.

of the Soviet population been officially revealed. In December 1986, for the first time in virtually a decade, the Soviet Union published age-specific death rates for its population. These numbers showed that between 1980–1981 and 1984–1985 mortality had declined for many age groups (a fact possibly related to the Gorbachev regime's decision to allow these data to be published). Even so, Soviet death rates had risen markedly from the levels of the 1960s. For persons in their thirties, death rates were up by about a sixth; for those over sixty, they had increased by about a quarter; for those in their forties and fifties, they had risen by half. For children under the age of five, death rates were said to be more than 11 percent higher in 1984–1985 than they had been in 1969–1970.[4]

The consequence of these changes was a shortening of life spans in the USSR. According to recently published official figures, life expectancy for men and women together fell from seventy years in 1971–1972 to sixty-nine years in 1985–1986.[5] These figures, however, may understate the actual magnitude of the drop. A recent U.S. Bureau of the Census evaluation of recent Soviet statistics suggests that life expectancy at birth in the USSR may have fallen by nearly a year for women and by almost two years for men between 1970 and 1985.[6] In 1970, however, Soviet life expectancy may already have been in decline. If the Census Bureau's new estimates are accurate, life expectancy in

TABLE 3–2
DECLINES IN LIFE EXPECTANCY AT BIRTH IN EASTERN EUROPE, 1964–1984

Country	*Period*	*Change in* E_0[a]	*Change in Male* E_0[a]	*Change in Female* E_0[a]
Bulgaria	1970–1980	–0.2	–0.7	+0.3
Czechoslovakia	1964–1983	–0.1	–0.9	+0.7
Hungary	1972–1983	–0.6	–1.8	+0.4
Poland	1975/78–1984	–0.3	–0.5	0.0
Romania	1976/78–1981	–0.2	–0.6	+0.2

a. E_0 is life expectancy at birth, in years.
SOURCE: Nick Eberstadt, *The Poverty of Communism* (New Brunswick, N.J.: Transaction Publishers, 1988), p. 215.

the USSR may have fallen since the mid-1960s by more than two years for women and by more than four years for men. For the Soviet population as a whole, life expectancy in the mid-1980s was apparently lower than it had been in the late 1950s.[7] Indeed, by the mid-1980s, life expectancy at birth was probably lower for a boy in the Soviet Union than for one in Mexico.

The spectacle of a pronounced and long-term deterioration in health for broad portions of the population of an industrialized country may be remarkable, but it is no longer unique. For as it happens, secular declines in health are now documented in every Communist country in Eastern Europe.

East Bloc States. The Warsaw Pact countries of Eastern Europe continued to publish detailed demographic statistics during the late 1970s and early 1980s and indeed issued figures until the demise of communism—and the pact—for their respective states. According to these data, life expectancy at birth has not only ceased to rise in recent years but actually fell slightly in five countries out of six (table 3–2), that is, in all countries of Soviet bloc Europe except East Germany.

These six East bloc states did in fact experience a decline in infant mortality rates during the 1970s and early 1980s although the decline proceeded at a distinctly slower pace than in Western Europe. But life spans for adults diminished in every one of these countries between the mid-1960s and the early 1980s. Some of the increases in adult mortality were dramatic: in Hungary, for example, death rates for persons in their forties rose by more than half between 1966 and 1982.[8] Eastern Europe's performance contrasts starkly with Western Europe's, where declines in adult mortality were not only continuing but accelerating (table 3–3).

TABLE 3–3
MORTALITY CHANGE FOR ADULTS IN THE USSR, EASTERN EUROPE, AND SELECTED WESTERN EUROPE NATO COUNTRIES, 1960S–1980S
(percent)

Age Group of Males	*USSR*	*Eastern Europe*[a]	*Western Europe*[a]
30–34	19	10	–25
35–39	20	22	–21
40–44	37	44	–18
45–49	29	50	–11
50–54	26	37	–10
55–59	22	28	–11
60–64	20	11	–13
65–69	25	9	–10
70–74	—	(11)	–7
75–79	27	(13)	–8
80–84	—	(9)	–5

Age Group of Females	*USSR*[b]	*Eastern Europe*[a]	*Eastern Europe*[c]	*Western Europe*[a]
30–34	–6	–12	–5	–31
35–39	–5	–14	–9	–31
40–44	5	–7	–2	–30
45–49	6	2	0	–26
50–54	13	3	2	–21
55–59	15	3	1	–20
60–64	4	–1	–3	–21
65–69	12	–3	–1	–26
70–74	—	(–4)	–5	–27
75–79	11	(–6)	–2	–26
80–84	—	(–3)	–4	–19

a. Mid-1960s to early 1980s.
b. Early 1960s to 1974.
c. 1972 to early 1980s.
— = not available.
( ) = average for available data.
SOURCE: Nick Eberstadt, *The Poverty of Communism* (New Brunswick, N.J.: Transaction Publishers, 1988), pp. 219–22.

In both the East and the West a number of arguments are currently offered to explain away the apparent deterioration of health conditions in Warsaw Pact countries. The first is that rising rates of mortality are only a statistical artifact, a reflection of improvements in vital registration, not real decreases in longevity. The argument is wholly unconvincing for Eastern Europe, where vital registration systems have been providing near-complete enumeration for decades. As for the USSR, apart from definitional questions surrounding the death of newborns, the analyses of both Soviet demographic authorities and the U.S. Census Bureau have concluded that mortality registration was virtually complete by the late 1960s[9] before most of the reported rise in age-specific death rates occurred. (In any event, reported increases in mortality in both Eastern Europe and the Soviet Union have been sharpest for middle-aged adults, precisely the group for whom demographic coverage is likely to be best.)

A second argument concedes that reported declines in health levels may be real but holds that they are simply the result of the stress attendant upon rapid modernization. If this argument were valid, one would have expected such societies as Japan and Taiwan to have suffered severe health reversals, whereas in actuality their pace of health progress over the past three decades has been both steady and rapid. Finally, it is sometimes argued that the health problems evidenced in the East bloc are a delayed consequence of shocks and stresses from World War II. Yet this would not explain why mortality for middle-aged men has risen in East Germany while it has been falling in West Germany, or why death rates have been rising for many groups born after 1945—apparently including Soviet infants born in the 1970s and 1980s.

What accounts for the pervasive deterioration in public health in the USSR and Eastern Europe in recent decades? Unfortunately, we know much less about the phenomenon than we might. Public discussion and scientific inquiry in these countries cannot proceed without the approval of the state. Although the local reversal in health progress is an issue of interest to many millions of persons in these countries, it is also a topic of extreme political sensitivity to their governments.

The proximate causes of increased mortality are probably diverse. They may include a rise in work-related accidents, a rise in tobacco and drug use, and increased pollution. (This latter phenomenon is taken quite seriously by local scientists and intellectuals. In Soviet Armenia, for example, one *samizdat* [self-published publication circulated unofficially] made the claim that "the number of mentally retarded children has risen fivefold; that of leukemia fourfold; and that of abnormal and premature births sevenfold" between 1970 and 1985.)[10]

As best as can be determined, however, two factors may have played a more important role than most. One is the rise in the consumption of alcoholic beverages, particularly in the consumption of hard liquor. Although the Eastern economies do not have a reputation for producing and supplying consumer goods efficiently, they have proved to be extremely effective in supplying their populations with alcoholic beverages. (These countries may also have a direct interest in encouraging their populations to imbibe: in the late 1970s Soviet figures indicated that taxes on alcohol amounted to more than 10 percent of the total revenues raised for the state budget.)[11] Between 1960 and 1980, per capita consumption of hard spirits in Eastern Europe appears to have nearly tripled.[12] In 1980 per capita consumption of hard liquor was estimated to have been almost twice as high in Eastern Europe as in Western Europe and more than three times as high in the USSR as in Western Europe. While one may speculate about the causes of their surge in drinking, its consequences—not only on the health of men and women but also on the health of their children—are beyond dispute.

The second factor in the rise in mortality in Eastern bloc countries may be the public health systems themselves. In the USSR the incidence of such diseases as influenza, measles, and scarlet fever are reported to be high and unremitting, and the incidence of diphtheria is reported to have risen substantially.[13] That such easily controllable infectious diseases should persist in an industrialized society would seem symptomatic of a basic failure of performance by the medical system.

Such failure may involve not only problems of administration but also problems of strategy. In the early days of Soviet power, the emphasis was upon turning out large numbers of medical personnel, not on training and equipping them. The health risks, and health needs, of the Soviet population changed, but it is not clear that the public health system made the corresponding adjustments. The ratio of doctors to population in 1980 was almost half again as high in Eastern Europe as in Western Europe and was more than twice as high in the USSR as in the twenty Western countries the World Bank lists as "industrial market economies."[14] Yet life expectancy at that time may have been as much as four years lower in Eastern Europe and as much as seven years lower in the USSR. Moreover health reversals in the Warsaw Pact countries became manifest precisely as their doctor-to-population ratios surpassed those of the West.

The Soviet bloc states seemed to view increased ratios of medical personnel to population not as a complement to greater financial commitments to medical care but rather as a substitute for it. Between 1965 and 1983 the percentage of public consumption funds allocated to public health and physical education in the Soviet Union *dropped,* from 16.5 to 14.2 percent. In the five Eastern Europe countries for which comparable figures are available, the average fell from 18.9 to 17.1 percent.[15] The Soviet Union and Eastern Europe may have been the only region of the world in which the share of national resources devoted to medical care fell in recent decades.

One may wonder why these socialist states were reducing the priority of health care at the very time that major public health problems were surfacing. For many of these countries, the 1970s and 1980s were a period of economic and budgetary difficulty. Socialist planning theory classifies health care as part of the "nonproductive sector"—a grouping of services that are held to contribute no value to the socialist economy, and thus become candidates for special scrutiny during times of budgetary stress. ("Nonproductive" persons may fall under the same cold scrutiny in the East bloc; exiles and émigrés have alleged that doctors and medical personnel in Czechoslovakia were regularly provided with an updated list of medicines that they were prohibited from prescribing to patients over age sixty-five unless these patients were of sufficient political stature.)[16]

Consumers of health services in Eastern Europe and the USSR could have little say about their government's decision to redirect resources away from medical care. The medical system in these countries was, at least in principle, almost fully socialized. The state did indeed pay directly for medical services used by the population, but it also decided what to provide and how much to spend.

People's Republic of China. In the People's Republic of China (PRC) substantial increases in longevity and reductions in infant mortality have been achieved since the advent of Communist power in 1949. By U.S. Census Bureau estimates (mortality registration in China remains incomplete), life expectancy at birth rose from about forty years in the early 1950s to about sixty-five in 1980.[17] This latter figure would compare quite favorably with contemporary estimates for India and Indonesia, two other populous, low-income Asian countries.

Even so, the Chinese performance with respect to health is not self-evidently superior. In Indonesia, India, and most other non-Communist Asian countries, government policy has sought to stabilize and to reduce the death rate, regardless of the pace of progress in mortality reduction. In the PRC, by contrast, the road to health progress has been anything but steady. In 1960, by U.S. Census Bureau estimates, China's life expectancy plunged to about twenty-five years, a drop of almost 50 percent in a matter of months.[18]

The reason for this collapse was not a natural disaster or war—such things are seldom sufficiently catastrophic to evince such an impact. Rather, it was a policy directive implemented nationwide at the instruction of the Chinese Communist party (CCP). In the late 1950s the CCP embraced the notion of a Great Leap Forward, a crash modernization program to lift China into the ranks of the fully industrialized countries by the early 1970s. In this utopian program of social engineering, the rural populace was forcibly reorganized into agricultural communes with enormous common kitchens and few personal or familial incentives for sustaining production. Inexperienced political operatives directed seasoned farmhands in the means by which harvests might be doubled or tripled, existing quotas were summarily increased upon signs that they might be achieved, and all extractable surplus from the countryside was sunk into industrial projects, many of them new and justified merely on the basis of enthusiasm. Although the adverse repercussions of this experiment were almost immediately evident in many localities, the campaign was nevertheless forced forward. The PRC lacked not only the mechanism by which politically sensitive information might be reliably transmitted up the party structure but also institutional restraints upon the absolute power of a directorate that was committed to the radical transformation of society.

In a short time, the Great Leap Forward succeeded in crippling the nation's industry and, perhaps even more importantly, in shattering the national food system. A period of desperate chaos and deprivation ensued. No one will ever know the exact human toll of the Great Leap Forward, but U.S. Census Bureau estimates suggest that the period following the campaign—to which the Chinese refer as the Three Lean Years—may have witnessed "excess mortality" approaching 30 million people.

While the Great Leap Forward is designated a terrible historical mistake by China's current party leadership, official criticism focuses upon tactics, not on questions of principle. The principle of purposely sacrificing particular human beings in the greater interest of the people (as determined by the party at any given time) has never been repudiated in China. Nor can it be repudiated by any Marxist-Leninist party, for to do so would be to attack the basis of the dictatorship of the proletariat by which such parties explain and justify their rule. Thus, although the party may well embrace programs or policies that confer material benefit upon the great majority of a population, there can be no guarantee that its policies may not simultaneously do grievous injury to given individuals or even to entire groups.

As recent events in China illustrate, this is not simply a hypothetical concern. Since the late 1970s, when the group surrounding Deng Xiaoping came to power, economic policies, particularly those regarding agriculture, have been readjusted. Food output (among other things) responded vigorously. At the same time, however, China codified and embraced a radical program of population control. Throughout the country political cadres relentlessly promoted the one-child norm—even to the point of establishing local quotas for births and of severely penalizing or punishing parents who violated them. One of the con-

TABLE 3–4

ESTIMATED INDEXES OF ECONOMIC AND DEMOGRAPHIC CHANGE IN THE PEOPLE'S REPUBLIC OF CHINA DURING THE "PERIOD OF READJUSTMENT," 1978–1982

Change	*1978*	*1982*	*%*
National income per capita (RMB)	315	421	+34 (+21)
Consumption per capita (RMB)	175	266	+52 (+32)
Grain consumption per capita (kg)	195.5	225.5	+15
Doctors per 10,000 population	10.8	12.9	+19
Infant mortality rate (deaths per 1,000)	37	46	+24
Life expectation at birth, both sexes (years)	65.1	64.7	−1

NOTE: National income figure refers to PRC conception of national income. Parenthetical figures indicate inflation-adjusted changes, as estimated by PRC State Statistical Bureau.

SOURCES: Lines 1–4: Derived from PRC State Statistical Bureau, *Statistical Yearbook of China 1983* (Hong Kong: Economic Information Agency, 1983), pp. 10, 22, 483. Lines 5–6: Judith Banister, "Analysis of Recent Data on the Population of China," *Population and Development Review*, vol. 10, no. 2 (June 1984), pp. 241–71.

sequences of this policy was the reemergence of infanticide in the Chinese countryside. Many parents apparently concluded that if they could have one child, it must be a boy. (By tradition sons, not daughters, support aged parents in rural China.) By the estimate of one American demographer, as many as a quarter of a million baby girls may have perished in connection with the population campaign in the early 1980s.[19] So pervasive was the policy-induced infanticide that it apparently pushed up China's infant mortality rate by nearly a quarter and caused total life expectancy to fall. The policies of the Chinese "readjustment" thus resulted in a peculiar contradiction: although by many material measures the well-being of the population seemed to be increasing, the health of the nation was apparently deteriorating (table 3–4).

Cuba. Cuba is often cited as a country that has made special strides toward better health under communism. Such claims—commonly accepted even by persons critical of the Castro regime—deserve further examination.

Contrary to Fidel Castro's assertion that "we had to start practically from zero," Cuba was one of the healthiest and most affluent tropical countries at the time he came to power. (In the early 1950s life expectancy in Cuba may have been higher than in such places as Spain, Portugal, and Greece.)[20]

According to Cuban data, over the 1960s life expectancy on the island may have risen by as much as six years—from about sixty-four years to about seventy years.[21] This would constitute a considerable improvement in public health. Yet other Caribbean and Latin American societies achieved similar (though less widely publicized) improvements over similar time spans. Among the examples are Guyana (seven years: 1950/1952—1959/1961), Costa Rica (seven years: 1962/1964—1972/1974), and Puerto Rico (seven years: 1949/1951—1954/1956).[22]

Cuban officials, and some foreign admirers, point with pride to Cuba's progress in combatting infant mortality. In 1985, according to official figures, Cuba's infant mortality rate was 16.5 per thousand live births. While this would represent a comparatively advanced level of infant health in the context of today's developing regions, it would not be unique. To the contrary it would be similar to the 1985 data that the U.S. Census Bureau reports for such Caribbean islands as Jamaica (18 per thousand), Martinique (16), Puerto Rico (15), Barbados (14), the Cayman Islands (13), and the Netherlands Antilles (11).[23]

Moreover there is reason to wonder whether Cuba's current level of infant health and its pace of progress in reducing infant mortality are as good as its figures suggest. Over the past decade and a half a number of curious inconsistencies have appeared in Cuba's demographic and public health statistics. The simplest explanation for these many inconsistencies would be that infant mortality data were being deliberately falsified.

Table 3–5 highlights one of these inconsistencies. Between 1970 and 1985 the reported incidence of several infectious diseases increased dramatically. Acute diarrhea is measured as having risen by more than a third, and acute respiratory diseases have reportedly more than tripled. Both of these conditions are strongly associated with infant mortality. Yet according to the Cuban figures infant mortality per thousand fell by almost three-fifths, from about 40 to under 17, during this period.

Other Cuban data point to similar contradictions. According to official numbers from the vital registration system, the infant mortality rate in Cuba dropped by one-fourth between 1970 and 1974. Yet according to life expectancy tables—also produced by the Cuban government—the infant mortality rate during this period actually rose by more than 11 percent.[24]

This sort of anomaly has not gone unnoticed by foreign demographers. According to an analysis sponsored by the National Academy of Sciences, revolutionary Cuba's demographic data are unusually good except for its data on infant mortality. The report says that "from the early 1970s onwards, consistency between indirect and official rates

TABLE 3–5
REPORTED INCIDENCE OF SELECTED DISEASES IN CUBA, 1970–1985
(incidence per 100,000 persons)

	Year		*Index*	
Disease	*1970*	*1985*	*1970*	*1985*
Acute diarrhea	7,628	10,487	100	137
Acute respiratory infection	10,162	38,160	100	376
Chicken pox	148.8	820.8	100	552
Diphtheria	0.1	—	100	—
Hepatitis	101.7	209.1	100	206
Malaria	—	4.5	NA	NA
Measles	104.3	28.5	100	27
Syphilis	7.2	62.6	100	869
Tetanus	2.6	0.1	100	4
Tuberculosis	30.5	6.7	100	22
Typhoid	4.9	0.6	100	12

NA = not applicable.
— = not reported.
SOURCE: *Anuario Estadistico de Cuba 1985* (Havana: Comite Estatal de Estadisticas, n.d.), pp. 307–11.

disappears: the indirect estimates indicate constant or even rising child mortality, while official figures show a continued rapid decline. . . . The sharp drop from the mid-1970s to 1980 is not supported by child survivorship data. . . . "[25]

No outside observer can ascertain definitively whether Cuban authorities have been falsifying infant mortality figures. The Cuban government, however, does have a record of falsifying particular statistics to which it ascribes political significance, as Fidel Castro explained it, to confuse "the enemies of the revolution." Moreover, on more than one occasion Cuban defectors have alleged that the Castro regime has falsified health figures. One former midlevel health official charges that the government suppressed information about a dengue fever epidemic in 1981, even to the point of doctoring death certificates.[26] And evidence has been accumulating to suggest that the Cuban government has been consciously understating the prevalence of AIDS (a disease that Cuban health authorities did not even acknowledge to be on the island as of January 1987.)[27] In a society where the government maintains a monopoly on official information and free speech is illegal, even serious public health problems may be concealed for a number of years.

Nutrition

Because of the close correspondence between health and nutritional well-being, the data on life expectancy and mortality in the preceding pages may also speak broadly to nutritional circumstance. In the Soviet Union, recently released data on the infant mortality rate put it at about 26 per 1,000 in 1985. This would be higher than the 1985 rates for such places as Chile, Malaysia, Panama, and Fiji.[28] But there is reason to believe that until at least the early 1970s the Soviet procedure for counting infant deaths was excluding as many as a fifth of the actual total.[29] As one might expect, a country with a reported infant mortality rate about three times that of the West may also have nutritional problems not typically associated with industrialized societies. In the 1970s one Soviet medical journal reported that 6 percent of the seven-year-olds surveyed in Leningrad suffered from rickets or hypertrophy. In some (though no means all) Western countries it might be possible to locate children with similar signs of malnutrition; the Soviet report, however, referred not to the hinterlands but to one of its showcase cities.

Famously unpalatable though the diet in the Soviet Union and Eastern Europe may have been, the caloric adequacy of the food supply in these regions is no longer in question. Nutritional problems in these countries in recent years seemed instead to relate to the problems of daily life under socialism. With the profound rise in drinking, and the feminization of alcoholism, child care—including infant feeding—may have suffered. The state's health monopoly evidently provided inadequate services for many afflictions, yet the vulnerable had no alternative. In the countryside, the workings of the planning mechanism may have routinely resulted in local shortages, a circumstance most likely to have affected precisely those who were most susceptible to nutritional problems.

Even if the planning mechanism (with its production quotas, rationed supplies, requisitioned produce, administratively imposed prices, and the like) were to work as in idealized description, it could not be expected unambiguously to improve the nutritional well-being of the population living under it. The purpose of the planning mechanism is to impose a single set of preferences—those determined by the party—over the activities of an economy that would otherwise respond to the preferences of individuals. Like hunger, nutritional well-being is in no small degree a subjective condition, a condition whose assessment depends directly on the assessment of those individuals whose judgments are purposely and systematically ignored through the planning mechanism.

TABLE 3–6
ESTIMATED FRACTION OF TOTAL CALORIC AVAILABILITY DERIVED FROM GRAIN IN THE PEOPLE'S REPUBLIC OF CHINA AND SELECTED OTHER AREAS, 1952–1981

	Period	*Total Caloric Availability*	*Percentage from Grain*
People's Republic of China	1952	1,917	88.3
	1957	2,065	88.4
	1960	1,462	87.0
	1970	2,092	89.5
	1975	2,226	89.0
	1979–81	2,531	87.1
Group 1			
Hong Kong	1972–74	2,596	48.4
Peninsular Malaysia	1972–74	2,428	62.3
Singapore	1972–74	2,787	49.8
Group 2			
Brazil	1972–74	2,471	52.9
Mexico	1972–74	2,625	58.6
Sri Lanka	1972–74	2,071	65.3
Thailand	1972–74	2,297	78.4
Group 3			
Bangladesh	1972–74	1,949	86.0
Ethiopia	1972–74	1,879	82.9
Java (poorest 40%)	1978	1,747	80.2

NOTE: Caloric availability given in PRC estimated calories per person per day. Grain, after standard official Chinese usage of the term, is taken to include not only cereals but roots and tubers, and pulses and legumes. All estimates but line 11 are from the food balancesheet approach; line 11 is derived from a nutrition survey. Caloric estimates to four places reflect the computation process, not implicit accuracy.
SOURCE: Nick Eberstadt, "Material Poverty in the PRC in International Perspective," *Issues and Studies*, vol. 22, no. 5 (May 1986).

Table 3–6 illustrates the tension. In recent years per capita caloric availability in China is thought to have risen substantially. Given a choice, most people prefer a varied and tasty diet. Nevertheless the composition of China's food supply seems to have changed little since the early 1950s. Once again we witness a paradox. To judge by caloric availability alone, the Chinese population would seem to have come near to the point of nutritional satiation in recent years. Yet to judge solely by their pattern of food consumption, one would assume that this was a population in dire poverty, if not one actually enduring famine.

Famine itself is not unknown to populations living under communism. The Great Leap Forward has been mentioned. Devastating famines have also struck in the Soviet Union, Mongolia, Vietnam, and more recently Cambodia and Ethiopia. Indeed, if a person died of famine after 1917, the odds are that he or she lived under a Marxist-Leninist government.

Famines in Communist countries have typically been the direct consequences of party and state policies, such as the forcible collectivization of the countryside. Often Communist governments have used famine as an instrument of social reconstruction and have fashioned hunger into a weapon against "enemies of the people." After North Vietnam's collectivization-induced famines had subsided, former landlords were still suffering from food shortages.[30] When an entire population depends upon rationed food for its sustenance, a government may apply hunger with precision. Since "enemies of the people" have no rights under communism—not even the right to live or to feed themselves—and since any person or group may eventually be designated as a hostile element, the nutritional status of the individual under communism in an important sense must always be provisional, for it can be affected at any time by a change in political climate.

Literacy

Figures on literacy must be used with caution. The definition of literacy is not absolute: it depends upon requirements that can vary between, and even within, societies and that tend to change over time. The evaluation of literacy moreover is a tricky business under the best of circumstances. It is made no easier by the techniques of quick, mass surveys.

Illiteracy statistics become all the more problematic when a government develops a political interest in seeing high literacy rates. When cadres are punished for failing to achieve mass education targets, and when a government indicates that it is prepared to be credulous about even extravagant claims of local success, measured literacy rates can rise rapidly even if the proportion of the population functionally literate undergoes little change.

Gerard Tongas, a French professor who taught in North Vietnam in the late 1950s, has given a valid picture of the anti-illiteracy campaign in that country.

The "fight against illiteracy"—insofar as it concerned the

> "elementary stage" at which the vast majority of the population was included—merely consisted in teaching the illiterate masses to recite twenty or so slogans and to copy them more or less legibly. "Long live President Ho!" "Long live the Vietnamese Workers' Party!" "Long live Peace!" "The Imperialism of the Americans and their lackeys will be defeated!" "Long live our Soviet comrades!" etc. Such were these slogans. The illiterates were to see them written thousands of times on walls, streamers, and in the newspapers; they would be able to recognize them, but never to read anything else![31]

Whatever purposes this approach to mass education may have served, enhancing the ability of the individuals to learn about the world would not seem to be among them.

Vietnam is not the only Communist state to have countenanced exaggeration of educational progress. North Korea and Cuba both announced, in the early stages of their revolutions, that they had completely eliminated illiteracy. In North Korea, according to official reports, the task was completed in only a few months. North Korea maintains to this day that it is a country of universal literacy, but over the past forty years it has never released the data against which its claim might be checked. For a while Cuba stood by the assertion that it had vanquished illiteracy; then it declared total victory against illiteracy a second time—releasing no documentation in either instance. When Cuba finally released its 1970 census results, the figures put the adult illiteracy rate at about 13 percent.[32]

Cuba's 1953 census has placed the nation's illiteracy rate at about 24 percent. Even if no progress had been made between 1953 and Castro's seizure of power in 1959, revolutionary Cuba's pace of illiteracy reduction, to judge by official figures, would have been little different from that of countries such as Chile, Costa Rica, and Panama. Indeed: the only Latin American or Caribbean country against which Cuba's performance appears unambiguously superior is Argentina.

In China, according to its 1982 census, the illiteracy rate for adults was about 35 percent, much lower than in India but no lower than in Indonesia (which may have begun the 1950s with a higher illiteracy rate than the mainland). China's census in 1982 suggests that adults over 55 years of age had an illiteracy rate above 75 percent.

Whatever educational results India and Indonesia may have achieved, they have gained through steady increments. The same cannot be said of China. Education in the PRC was marked by a period known as the Cultural Revolution. In the late 1960s and early 1970s the entire schooling system was deliberately disrupted, and for a time suspended, because the party leadership accused it of being dangerously "bourgeois." Students were turned out into the streets and encouraged to organize into gangs of Red Guards to confront and attack persons and institutions suspected of opposing Party Chairman Mao Zedong. At the height of the turmoil that ensued, hundreds of thousands of persons are thought to have lost their lives.

China's rulers today refer to the Cultural Revolution as a "lost decade." Indeed, whatever the problems may be with education in India, a child raised in that caste-bound country stood a far better chance of going to high school or college in the 1960s and 1970s than did one in China.

An Irresolvable Contradiction

Upon examination, the record of the past seventy years does not lend support to the argument that Communist governments, or their policies, achieve unambiguously or even generally superior results in alleviating material poverty. Communist governments are indeed committed to a radical transformation of the societies beneath them, and Marxist-Leninist doctrine requires the concentration and augmentation of state power (under the direction of the party) for this purpose. To the extent that strategies for augmentation of state power coincide with improvements in individual well-being, Marxist-Leninist policies may result in reductions in material deprivation. But there can be no assurance that the quest for state power will necessarily improve the material well-being or security of the population affected, or that it will not instead result in terrible and unpredicted hardships.

The distinction between purposes of state and the well-being of the individual is vital to an understanding of poverty under communism. Poverty is irreducibly a phenomenon experienced by human beings. Correspondingly, alleviating poverty involves the extension of human choice. But Communist rule is programmatically hostile to precisely those institutions that articulate and secure individual choice: private property, free markets, constitutional liberties, rule of law. Communist governance aggressively attacks these institutions in principle, for they threaten the basic Leninist conception of unrestrained state power. Instead of such "bourgeois rights," Communist states propose the notion of "people's rights"—for which their party leadership will be the ultimate, and unrestrained, interpreter. Unfortunately, while enshrining the concept of people's rights, these parties do not recognize as legitimate the proposition that these rights may be divided into individual portions. The performance of Communist states in providing for the poor within their boundaries is shaped, and ultimately limited, by this irresolvable contradiction.

4

The Decline of Public Health in Eastern Europe, 1965–1985

Students of public health and informed nonspecialists alike are by now generally aware that the Soviet Union suffered a pronounced and protracted deterioration in health for much of its population over the past several decades. The outlines of this health problem were first noted in the mid-1970s.[1] By the early 1980s the phenomenon had become a topical focus of analysis and commentary in the West.[2] By the late 1980s, several years into the Gorbachev glasnost campaign, Soviet officials were publicly examining, and declaiming upon, the dimensions and the causes of the health setbacks the country had suffered over the preceding generation.[3]

The countries of Communist Eastern Europe also were beset by mounting health problems over the past generation. In a variety of respects, these problems parallel those of the contemporary USSR. To be sure, Eastern European health setbacks on the whole have been less marked. Nevertheless throughout Eastern Europe[4] death rates for adult age groups have registered long-term increases. Declining life expectancy at birth, moreover, has been a characteristic trend for the region over the past generation.

Glasnost notwithstanding, official discussion of local health problems has been (and remains) more open, and health-related data more comprehensive, for Eastern Europe than for the Soviet Union. Such data and discussions provide a somewhat more detailed picture of the anatomy of secular health decline in industrial society than are to date available from the USSR.

TABLE 4–1
DECLINES IN LIFE EXPECTANCY AT BIRTH IN EASTERN EUROPE AND THE USSR, 1964–1985

Country	*Period*	*Change in* E_0 *(years)*	*Change in Male* E_0	*Change in Female* E_0
Bulgaria	1970–80	–0.2	–0.7	+0.3
Czechoslovakia	1964–83	–0.1	–0.9	+0.7
GDR	1967/68–1976	–0.1	–0.3	+0.0
Hungary	1972–85	–0.6	–1.8	+0.5
Poland	1974–85	–0.6	–1.3	+0.2
Romania	1976/78–1982/84	–0.1	–0.6	+0.4
Yugoslavia	1979/80–1984/85	–0.1	–0.6	+0.4
USSR	1964/65–1984	–2.7	–3.7	–1.2

E_0 = life expectancy at birth.
SOURCES: Bulgaria: United Nations, *World Population Trends, Population and Development Interrelation and Population Policies: 1983 Monitoring Report*, vol. 1 (New York: UN, 1985). Czechoslovakia: United Nations, *Demographic Yearbook 1969* (New York: UN, 1970); *Statisticka Rocenka 1985* (Prague: Federainy Statisticky Urad, 1985). German Democratic Republic: United Nations, *Demographic Yearbook, Special Issue, Historical Supplement* (New York: UN, 1979). Hungary: *Demografiai Evkonyv 1986* (Budapest: Kozponti Statisztikai Hivatal, 1987). Poland: United Nations, *Demographic Yearbook 1986* (New York: UN, 1988). Romania: *Anuarul Statistic al Republicii Socialiste Romania 1986* (Bucharest: Directia Centrala de Statistica, 1987). Yugoslavia: *Demografska Statistika 1985* (Belgrade: Savezni Zavod Za Statistiku, 1988). USSR: Anatoliy Vishnevskiy, "Has the Ice Cracked? Demographic Processes and Social Policy," *Kommunist*, no. 6 (April 1988), pp. 65–75, translated in Joint Publications Research Service, series UKO, no. 88-011 (July 11, 1988), p. 44.

Dimensions of Mortality Change

Indicators of health and disease for human populations are diverse. But as K. Uemura has noted, "data on mortality are the most standardized of all disease statistics."[5] The singular import of this particular measure of health, moreover, is beyond dispute.

The simplest, and most intuitively obvious, indicator of total mortality for a national population is its expectation of life at birth. Table 4–1 presents data on life expectancy at birth for Eastern Europe and the USSR. In the 1950s and early 1960s the pace of improvement in life expectancy at birth in the Soviet Union and Eastern Europe had been quite rapid. The United Nations Population Division estimates that between the early 1950s and the early 1960s life expectancy at birth rose more than five and a half years for the countries of Soviet Bloc Europe and six and a half years in Yugoslavia; by contrast the increase for the United States during that same period is estimated to have been only one year.[6] By the mid-1960s, however, improvements in life expectancy in Eastern Europe had decelerated sharply, even as progress in life expectancy in Western Europe and North America was quickening. In the 1970s and early 1980s, increases in life expectancy for women in Eastern European countries were at best halting, and life expectancy at birth for men fell throughout the region. Life expectancy at birth generally ceased its rise; indeed, every country in Eastern Europe has registered at least some decline in this measure in the period since 1965. With the single exception of the then–German Democratic Republic, these declines have continued through the recent period.

The drop in life expectancy at birth in the various countries of Communist Europe has not been nearly as sharp over the past generation as in the USSR. Even so, Eastern Europe's recent health record seems to represent something fundamentally new. In the past, industrialized countries have witnessed periods of slow health progress (as 1955–1965 proved to be for many OECD member states). Many industrialized countries, moreover, have registered slight, temporary declines in life expectancy on their advance to greater longevity. No region of the industrialized world, however, has heretofore experienced the sort of interruption, and actual reversal, of health progress during peacetime that was recorded in the societies of Communist Europe.

Life expectancy at birth is a summary measure reflecting survival probabilities for individuals of all ages. To understand what occurred in Eastern Europe, it is useful to separate this measure into two subsidiary components: survival chances for children from birth to the age of one (as represented in the infant mortality rate) and expectation of life at one year of age.

Unlike the USSR (where, after a decade of statistical silence, authorities finally reported the infant mortality rate for the mid-1980s to be higher than the one recorded in 1970), infant mortality rates in Eastern Europe underwent steady decline between 1965 and 1985. The tempo of improvement in Eastern European infant mortality rates, however, was slower during those years than in Western Europe—although infant mortality rates for Western Europe as a whole were already considerably lower (table 4–2).

In 1985 infant mortality was reported to be roughly two-thirds higher in Soviet Bloc Europe than in Western Europe. These numbers, however, may actually understate the true differential. As A. Klinger noted in 1982,

TABLE 4–2
RECORDED INFANT MORTALITY RATES IN EASTERN AND WESTERN EUROPEAN COUNTRIES, 1960–1985
(deaths per 1,000 live births)

Country or Group	*1960*	*1965*	*1975*	*1980*	*1985*
Bulgaria	45	31	23	20	15
Czechoslovakia	24	26	21	17	15
GDR	39	25	16	12	9
Hungary	48	39	33	23	20
Poland	56	42	25	21	18
Romania	77	44	35	29	23
Unweighted average, Eastern Europe	48	34	25	20	17
Unweighted average, 18 Western European countries	31	25	16	12	10
Ratio, Eastern Europe to Western Europe	1.54	1.36	1.56	1.73	1.69
	1960–65	*1965–75*	*1975–85*	*1960–85*	
Percent of Decline in Recorded Mortality Rates					
Unweighted average, Eastern Europe	–29	–26	–34	–65	
Unweighted average, 18 Western European countries	–19	–35	–39	–72	

NOTE: Figures are presented only to two places and thus may not add or average because of rounding.
Eighteen Western European countries: Austria, Belgium, Denmark, Federal Republic of Germany, Finland, France, Greece, Iceland, Ireland, Italy, Luxembourg, Netherlands, Norway, Portugal, Spain, Sweden, Switzerland, United Kingdom, with the following exceptions: 1975 does not include Iceland; 1980 does not include Switzerland; 1985 does not include Iceland or Luxembourg.
SOURCES: 1960–1980, Eastern Europe, 1960–1985 Western Europe: World Bank, *World Tables*, vol. 2, 3rd and 4th ed. (Washington, D.C.: World Bank, 1983 and 1987); 1985, Eastern Europe: United Nations, *Demographic Yearbook 1986* (New York: UN, 1988).

> only a few countries in [socialist Europe] used the standard definitions [for infant mortality] provided by the UN and the WHO [World Health Organization] . . . even in 1979 national definitions in four of these countries—Bulgaria, Romania, and

to a lesser extent Poland and Yugoslavia—differed from the international recommendations.[7]

In Poland, according to M. Okolski, "the infant mortality rate is underestimated and [has been] artificially lowered since 1964."[8] By his estimate Poland's infant mortality rate in 1980 would have been almost a fourth higher than was reported if international WHO definitions had been used.[9] In Bulgaria and Yugoslavia, the proportion of total infant deaths attributed to the neonatal period (the first twenty-eight days of life) is strangely low, and inconsistent with the ratios reported by other European societies adhering to the international standard definition of infant mortality. Adjustments for the underreporting of neonatal mortality could raise the infant mortality rate in these two countries by 40 percent or more. As for Romania, reports of local practices indicate that births need not be registered at all during the first month—precisely the time when infant fatality is most likely. Perhaps not surprisingly, Romania reports that neonatal deaths account for only a small fraction of the country's infant mortality.[10] If its definitions and procedures for recording infant mortality conformed with the international norm proposed by the WHO, Romania's infant mortality rate might easily be 80 percent higher than what Bucharest reported.[11] (Such serious underreporting of infant mortality would have a consequential impact on total estimates of life expectancy; increasing Romania's measured infant mortality rate 80 percent would reduce the country's measured life expectancy at birth more than a year).

Definition and registration of death tend to be more uniform for children, youths, and adults than it is for babies. Table 4–3 presents data on life expectancy at age one in Eastern Europe, Western Europe, and the Soviet Union. Although East Germany enjoyed some improvement by this measure between 1965 and 1985, the rest of the Warsaw Pact did not. In fact the unweighted average for the group showed a decline for those years; for 1966 to 1985, life expectancy at age one fell nearly a year. While this drop was not as great as the USSR's, it contrasts with a rise of almost three years for an unweighted average of eighteen Western European countries. In Yugoslavia, life expectancy at age one did rise between the mid-1960s and the mid-1980s; improvements in non-Communist countries in Southern Europe (Greece, Portugal, and Spain), however, were more substantial.

Slow progress, or absolute decline, in life expectancy at age one in Communist Europe was caused principally by changes in life expectancy among adults. Table 4–4 depicts changes in life expectancy at age thirty in Eastern Europe, Western Europe, and the USSR. For Warsaw Pact Europe as a whole, life expectancy for women increased

TABLE 4–3
LIFE EXPECTANCIES AT AGE ONE IN EASTERN AND WESTERN EUROPEAN COUNTRIES, 1960–1985

Country or Group	*1960*	*1965*	*1966*	*1970*	*1975*	*1980*	*1985*[a]
Bulgaria	71.7	72.2	72.3	72.3	71.7	71.7	71.2
Czechoslovakia	71.2	71.0	71.1	70.2	70.9	70.7	71.1
GDR	70.7	71.3	71.4	71.0	71.4	71.2	72.2
Hungary	70.9	71.4	72.2	71.2	71.1	70.2	69.6
Poland	70.6	71.5	71.9	71.7	71.9	70.7	71.1
Romania	70.3	71.0	71.3	70.6	71.3	70.4	70.2
Unweighted average, Eastern Europe	70.9	71.4	71.7	71.2	71.4	70.8	70.9
Unweighted average, 18 Western European Countries	71.6	71.8	71.9	72.3	72.9	73.9	74.7
Yugoslavia	67.2	69.9	71.1	70.3	71.1	71.2	72.0
Unweighted average, Greece, Portugal, Spain	70.5	70.9	70.9	72.2	72.5	73.1	74.7
USSR	71.4	71.4	71.6	70.0	69.5	69.3	69.8[b]

a. Figures are for 1985 or most recent year available. Eighteen Western European countries: Austria, Belgium, Denmark, Federal Republic of Germany, Finland, France, Greece, Iceland, Ireland, Italy, Luxembourg, Netherlands, Norway, Portugal, Spain, Sweden, Switzerland, United Kingdom (England and Wales).
b. 1985–1986 figure.
SOURCES: 1960–1980: Jean Bourgeois-Pichat, "Mortality Trends in Industrialized Countries," in United Nations, ed., *Mortality and Health Policy* (New York: United Nations, 1984). 1985, Eastern Europe: Unpublished life tables prepared by the U.S. Census Bureau, Center for International Research, based on official mortality data by age and sex, 1988. 1985, Western Europe: Council of Europe, *Recent Demographic Developments in the Member States of the Council of Europe* (Strasbourg: Council of Europe, 1987), and European Economic Community, *Demographic Statistics 1988* (Brussels: Statistical Office of the European Community, 1988). 1985, Soviet Union: *Naseleniye SSSR 1987* (Moscow: Goskomstat, 1988).

only marginally between 1965 and 1985; it fell for men in every country, and by an unweighted average of more than two years. In each country, total life expectancy for persons thirty years of age was lower in 1985

TABLE 4–4
LIFE EXPECTANCIES AT AGE THIRTY IN EASTERN EUROPE, USSR, AND WESTERN EUROPEAN COUNTRIES, MID-1960S AND MID-1980S

		Life Expectancy (years)		*Change (years)*	
Country	Period	Male	Female	Male	Female
Bulgaria	1965–67	43.06	45.99		
	1985	40.6	46.1	–2.5	+0.1
Czechoslovakia	1964	41.15	45.84		
	1984	39.50	46.27	–1.7	+0.4
GDR	1967–68	42.46	46.70		
	1985	41.56	46.76	–0.9	+0.1
Hungary	1964	41.74	45.45		
	1985	38.38	45.61	–3.4	+0.2
Poland	1965–66	41.68	46.46		
	1985	39.21	46.65	–2.5	+0.2
Romania	1966	42.4	45.6		
	1985	40.2	45.3	–2.2	–0.3
Unweighted average, Eastern Europe	c. 1965	42.08	46.01		
	c. 1985	39.91	46.12	–2.2	+0.1
Unweighted average, Western Europe[a]	c. 1965	42.01	46.65		
	c. 1985	43.58	49.54	+1.6	+2.9
Yugoslavia	1966	42.5	46.0		
	1980–81	41.41	46.34	–1.2	+0.3
Selected Southern Europe (Greece, Portugal, Spain)	c. 1965	42.38	46.33		
	c. 1980	44.05	48.89	+1.7	+2.6
FRG	1965	41.21	46.03		
	1983–85	43.05	49.07	+1.9	+3.0
USSR	1965	45[b]			
	1985	42[b]			

a. Western Europe: Austria, Denmark, Federal Republic of Germany, Finland, France, Iceland, Ireland, Italy, Netherlands, Norway, Portugal, Sweden, Switzerland, United Kingdom (England and Wales).

(Notes continue on next page)

than in 1965. Once again, deteriorations in health conditions were less pronounced than in the USSR (where life expectancy for adult women almost certainly declined) but compared unfavorably with the health progress registered in Western Europe. Adult life expectancy in 1985, by this measure, was about three and a half years lower for both men and women in Soviet Bloc Europe than in Western Europe; in 1965 the levels had been virtually even. In Yugoslavia life expectancy for adult men declined between 1965 and 1985; total life expectancy for adults was also down. By contrast the average for Greece, Portugal, and Spain rose almost two years for men, and more than two years for women.

Age-specific mortality rates provide a more detailed glimpse at the changing patterns of adult health in Eastern Europe (table 4–5). Soviet Bloc Europe saw a broad rise in death rates for adult men between 1965 and 1985; only in East Germany were declines recorded, and there only for a few cohorts. For the region as a whole, death rates for men in their sixties rose about 10 percent between 1965 and 1985; for those in their thirties, more than 20 percent; for those in their fifties, more than 30 percent; for those in their forties, more than half.

Among women, rising death rates were registered for at least some adult cohorts in all Soviet Bloc countries, with the exceptions of East Germany and Romania. (Romania's data, however, are not for 1985. Between its 1986 and its 1987 editions, Bucharest's official statistical yearbook, *Anuaral Statistic,* collapsed from about 400 pages to scarcely 130 pages; mortality data were one of many topics omitted in the newly slim volume.) In Yugoslavia, mortality decline among adult women was more regular and substantial than in the rest of Communist Europe, but death rates for men in their forties, fifties, and early sixties rose measurably over the two decades.

What can account for this deterioration of health conditions among Eastern Europe's adult populations? Some have suggested that rising adult mortality may be a delayed consequence of the calamitous stresses suffered by local populations during World War II.[12] While this hypothesis has its place, it is inadequate to explain the peculiar

---

(Continued from preceding page)
b. Figure for both sexes.
SOURCES: Soviet Union: 1965, *Narodnoye Khozyaystvos SSR v. 1965 g.* (Moscow: Moskva Tsentralnoe Pri Sovete Ministrove USSR, 1966); 1985, *Vestnik Statistiki,* no. 3 (1987), p. 79. Bulgaria (1985), Romania (1985), Yugoslavia (1965): Unpublished life tables prepared by U.S. Census Bureau, Center for International Research, from official data on population and mortality by age and sex, 1988. All other data: United Nations, *Demographic Yearbook,* various years (New York: UN).

TABLE 4–5
CHANGES IN MORTALITY RATES FOR ADULTS IN EASTERN EUROPE, 1965–1985
(percent)

	Age								
Country	30–34	35–39	40–49	45–49	50–54	55–59	60–64	65–69	70–74
Bulgaria, 1965–85									
Males	+25	+38	+48	+67	+45	+44	+29	+19	+17
Females	–20	–14	–5	–7	–6	+5	+1	–7	–3
Czechoslovakia, 1965–84									
Males	0	+12	+24	+44	+38	+28	+17	+3	+6
Females	–13	–15	–14	–6	0	+1	+7	–3	–8
GDR, 1965–85									
Males	0	–5	+6	+14	+11	0	–6	–10	0
Females	–36	–25	–25	–20	–13	–11	–5	–13	–10
Hungary, 1965–85									
Males	+40	+69	+100	+118	+79	+58	+32	+12	+8
Females	+20	+27	+26	+26	+27	+14	+4	–8	–11
Poland, 1965–85									
Males	+14	+21	+46	+56	+51	+36	+21	+6	NA
Females	–36	–19	0	–3	+2	+1	–1	–10	NA
Romania, 1965–84									
Males	+14	+28	+50	+59	+35	+23	+2	+1	0
Females	–8	–12	–8	–5	–3	–7	–8	–9	–14
Unweighted average, Eastern Europe									
Males	+16	+27	+46	+60	+43	+32	+16	+5	(+6)
Females	–16	–10	–4	–3	+1	+1	0	–8	(–9)
Yugoslavia, 1965–85									
Males	–19	–4	+5	+20	+26	+13	+3	–7	–4
Females	–53	–37	–21	–19	–17	–13	–15	–23	–18

NA = not available.
( ) = unweighted average for countries with available data.
SOURCES: Yugoslavia: Compiled by U.S. Census Bureau, Center for International Research, from *Demografska Statistika (Belgrade)*, various issues. Other data: United Nations, *Demographic Yearbook 1974* (New York: UN, 1975); United Nations, *Demographic Yearbook 1986* (New York: UN, 1988).

health trends in evidence in Eastern Europe in their entirety. For one thing, death rates in Warsaw Pact Europe were typically higher in 1985 than twenty years ago for men in their early thirties—higher, in other words, for those born a decade after World War II than for those

TABLE 4–6
CHANGES IN MORTALITY RATES FOR ADULTS IN THE GERMAN DEMOCRATIC REPUBLIC AND THE FEDERAL REPUBLIC OF GERMANY, 1965–1985
(percent)

	Age								
Country	30–34	35–39	40–44	45–49	50–54	55–59	60–64	65–69	70–74
GDR									
Males	0	–5	+6	+14	+11	0	–6	–10	0
Females	–36	–25	–25	–20	–13	–11	–5	–13	–10
FRG									
Males	–28	–26	–17	–9	–14	–20	–25	–26	–18
Females	–40	–33	–35	–29	–28	–25	–31	–34	–34
Difference (GDR-FRG)									
Males	+28	+21	+23	+23	+25	+20	+19	+16	+18
Females	+4	+8	+10	+9	+15	+14	+26	+21	+24

SOURCES: United Nations, *Demographic Yearbook 1967, 1974,* and *1986* (New York: United Nations, 1968, 1975, and 1988, respectively).

who lived through it. Second, this rise in mortality in Eastern Europe is inconsistent with patterns of mortality described in research on the demographic after-effects of major conflicts. S. Horiuchi[13] has presented evidence that unusually high mortality in later life for those surviving a major war is characteristic of the cohort of boys just under draft age at the time of the war in question; in Warsaw Pact Europe, by contrast, the rise in mortality was general among adult males, and in such places as Hungary, general among females as well. Third, deterioration in adult health levels is not characteristic of all populations that suffered heavily during World War II. Japan suffered severe privation during and immediately after the Second World War, yet it has enjoyed substantial and steady improvements in adult health during the postwar decades. Within the German population—which experienced World War II as a single country—dramatic differences in adult health progress were apparent in 1985 between Eastern and Western Germany in every cohort (table 4–6).

Changing Cause-of-Death Patterns

How then are the health problems of Eastern Europe, circa 1985, to be explained? Some preliminary insights may be afforded by data on causes of death. Such data must be used with caution. Even in industrialized countries these figures are less standardized than one might suppose. As Z. Brzezinski has warned, "variations between different countries in diagnostic practices and coding of the death certificates give cause to doubt the validity of causes of death."[14] Even within the European Economic Community, he noted, "large differences in coded cause of death were found within and between countries."[15] Practices in certain Eastern European countries, moreover, are somewhat less than standard on their face: East Germany, for example, did not report deaths from homicide, suicide, or "accidents and adverse effects." For all these limitations, a review of recent data on mortality by cause of death may nevertheless prove instructive.

Tables 4–7 and 4–8 present data on age-standardized death rates by reported cause of death for Eastern and Western Europe. The reference population against which death rates are standardized, a European model devised by the World Health Organization is, like Europe itself, weighted toward older age groups in its composition. It therefore tends to be more sensitive to changes in mortality among the middle-aged and the elderly—precisely the groups that seemed to have suffered setbacks in health in Eastern Europe in recent decades.

By the mid-1980s age-standardized death rates for men were more than one-third higher in Warsaw Pact Europe than in Western Europe and more than two-fifths higher for women (table 4–7). For most (though not all) reported causes of death, standardized mortality rates for men and women were higher in Eastern than in Western Europe. In relative terms the greatest differentials were found in death from liver disease (including cirrhosis) and in diseases of the circulatory system (including heart attack, stroke, and arteriosclerosis). In absolute terms deaths from diseases of the circulatory system dominate the contemporary differential in standardized mortality between Eastern and Western Europe. More than three-fifths of the difference in standardized rates for men, and almost nine-tenths of the total difference for women, can be ascribed to differences in death rates from cardiovascular disease alone. In specific comparisons of more selected areas of then-Communist and non-Communist Europe (for example East versus West Germany; Yugoslavia versus Greece, Portugal, and Spain), standardized death rates from cardiovascular disease are consistently reported to be substantially higher in the East and can account for the great majority of existing mortality differentials for both males and females (tables 4–8 and 4–9).

Table 4–10 traces standardized mortality rates by reported cause of death back, where data permit, to the late 1950s. Absolute differences in age-standardized mortality between Warsaw Pact Europe and Western Europe narrowed slightly between 1955–1959 and 1965–1969 but have

TABLE 4–7

AGE-STANDARDIZED DEATH RATES FOR SELECTED CAUSES IN EASTERN AND WESTERN EUROPE, CIRCA 1985 (EUROPEAN MODEL)

Cause of Death	*Death Rate per 100,000*		*Absolute Difference (East-West)*	*Relative Difference (West = 100)*	*Cause as Proportion of Difference (%)*
	Eastern Europe	Western Europe			
All causes: male	1,507.1	1,110.0	397.1	136	
female	960.5	665.2	295.3	144	
Infectious, parasitic: male	10.7	8.4	2.3	127	1
female	4.5	4.6	–0.1	98	–1
Malignant neoplasms: male	260.2	261.8	–1.6	99	–1
female	149.5	154.1	–4.6	97	–2
Neoplasms of trachea, bronchus, lung: male	80.1	74.1	6.0	108	2
female	11.2	13.5	–2.3	83	–1
Circulatory system: male	801.3	500.1	301.2	160	76
female	573.6	310.6	263.0	185	89
Ischaemic heart: male	315.5	243.3	72.2	130	18
female	149.2	108.4	40.8	138	14
Respiratory system: male	124.7	96.6	28.1	129	7
female	59.7	47.5	12.2	126	4
Digestive system: male	68.4[a]	46.1	22.3	148	7
female	33.0[a]	25.3	7.7	130	3
Liver disease, cirrhosis:	35.6	21.7	13.9	164	4
female	13.6	8.0	5.6	170	2
Injury, poisoning: male	120.7[a]	81.5	39.2	148	10
female	45.0[a]	34.0	11.0	132	4
Traffic accidents: male	20.0[b]	22.7	–2.7	88	–1
female	5.8[b]	7.2	–1.4	81	–1
Suicide: male	36.9[a, b]	22.6	14.3	163	4
female	11.3[a, b]	8.9	2.4	127	1

NOTES: "Age-standardized death rates (European Model)" refers to the application of age-specific death rates to "European Model" population structure used by the World Health Organization.

Eastern Europe: Bulgaria, Czechoslovakia, German Democratic Republic, Hungary, Poland, Romania. Western Europe: Austria, Belgium, Denmark, Federal Republic of Germany, Finland, France, Greece, Iceland, Ireland, Italy, Luxembourg, Netherlands, Norway, Portugal, Spain, Sweden, Switzerland, United Kingdom (England and Wales).

a. Figure does not include GDR.

b. Figure does not include Romania.

SOURCE: Derived from World Health Organization, *World Health Statistics Annual 1987* (Geneva: WHO, 1987).

TABLE 4–8

AGE-STANDARDIZED DEATH RATES FOR SELECTED CAUSES IN THE GERMAN DEMOCRATIC REPUBLIC AND THE FEDERAL REPUBLIC OF GERMANY, CIRCA 1985 (EUROPEAN MODEL)

	Death rate per 100,000		*Absolute Difference (GDR-FRG)*	*Relative Difference (FRG = 100)*	*Cause as Proportion of Difference (%)*
Cause of Death	GDR	FRG			
All causes: male	1,399.9	1,136.6	263.3	123.0	
female	905.9	673.8	232.1	134.0	
Infectious, parasitic: male	6.2	8.8	–2.6	70	–1
female	3.4	4.8	–1.4	71	–1
Malignant neoplasms: male	243.3	275.0	–31.7	88	–12
female	148.7	166.5	–17.8	89	–8
Neoplasms of trachea, bronchus, lung: male	75.1	72.5	2.6	104	1
female	8.2	10.6	–2.4	77	–1
Circulatory system: male	751.8	528.8	223.0	142	85
female	534.9	326.9	208.0	164	90
Ischaemic heart disease: male	226.0	242.3	–16.3	93	–6
female	109.0	107.5	1.5	101	1
Respiratory system: male	115.4	91.0	24.4	127	9
female	41.3	35.1	6.2	118	3
Digestive system: male	NA	56.0			
female	NA	29.6			
Liver disease, cirrhosis: male	23.2	29.9	–6.7	78	–3
female	9.0	11.4	–2.4	79	–1
Injury, poisoning: male	NA	67.1			
female	NA	31.4			
Traffic accidents: male	15.5	18.7	–3.2	83	–1
female	5.2	6.7	–1.5	78	–1
Suicide: male	NA	25.1			
female	NA	10.1			

NA = not available.
SOURCE: See table 4–7.

widened rapidly since then. Differences in mortality from cardiovascular disease explain much of the total difference in standardized mortality between these two regions over the past two decades and indeed seem to help account for their disparate trends in general mortality.

TABLE 4–9
AGE-STANDARDIZED DEATH RATES FOR SELECTED CAUSES IN YUGOSLAVIA AND SELECTED SOUTHERN EUROPEAN COUNTRIES (GREECE, PORTUGAL, SPAIN), CIRCA 1985 (EUROPEAN MODEL)

Cause of Death	*Death Rate per 100,000*		*Absolute Difference (Yug-S. Eur.)*	*Relative Difference (S. Eur. = 100)*	*Cause as Proportion of Difference (%)*
	Yugoslavia	Southern Europe			
All causes: male	1,439.6	1,058.6	381.0	136	
female	978.8	683.7	295.1	143	
Infectious, parasitic: male	22.1	11.9	10.2	186	3
female	13.0	5.6	7.4	232	3
Malignant neoplasms: male	213.7	214.7	–1.0	100	1
female	122.6	118.4	4.2	104	1
Neoplasms of trachea, bronchus, lung: male	60.6	53.3	7.3	114	2
female	9.3	7.3	2.0	127	1
Circulatory system: male	713.7	444.8	268.9	160	71
female	559.3	331.7	227.6	169	77
Ischaemic heart disease: male	119.3	117.3	2.0	102	1
female	56.8	52.6	4.2	108	1
Respiratory system: male	101.3	87.4	13.9	116	4
female	60.5	45.5	15.0	133	5
Digestive system: male	73.6	59.8	13.8	123	4
female	31.7	26.5	5.2	120	2
Liver disease, cirrhosis: male	44.5	33.1	11.4	134	3
female	14.8	11.1	3.7	133	1
Injury, poisoning: male	97.6	79.6	18.0	123	5
female	34.3	27.7	6.6	124	2
Traffic accidents: male	32.3	31.8	0.5	102	1
female	8.4	9.0	–0.6	93	–1
Suicide: male	26.1	9.4	16.7	278	4
female	10.1	3.2	6.9	316	2

SOURCES: Yugoslavia: World Health Organization, *World Health Statistics Annual 1986* (Geneva: WHO, 1986). Other figures: Derived from World Health Organization, *World Health Statistics Annual 1987* (Geneva: WHO, 1987).

Whereas the level of deaths attributed to cardiovascular disease has been declining since at least the late 1950s for Western European men, and since the late 1960s for Western European women, in Eastern Europe the level has been on the rise for men since at least the late

TABLE 4–10
AGE-STANDARDIZED DEATH RATES FOR SELECTED CAUSES IN EASTERN AND WESTERN EUROPE, 1955–59 TO 1975–79 (EUROPEAN MODEL)
(per 100,000)

	1955–59		*1965–69*		*1975–79*	
Cause of Death	Male	Female	Male	Female	Male	Female
All causes						
Eastern Europe[a]	1,471.1	1,136.2	1,389.3	975.7	1,442.1	965.9
Western Europe[b]	1,400.8	1,030.4	1,344.9	903.2	1,238.0	761.2
Absolute difference (East-West)	70.3	105.8	44.4	72.5	204.1	204.7
Relative difference (West=100)	105	110	103	108	116	127
Circulatory system						
Eastern Europe	603.2	525.9	621.2	500.7	719.8	544.2
Western Europe	589.3	468.0	601.9	425.7	566.5	360.9
Absolute difference	13.9	57.9	19.3	75.0	153.3	183.3
Relative difference	102	112	103	118	127	151
Cause as proportion of difference (%)	20	55	43	103	75	90
Respiratory system						
Eastern Europe	147.9	101.1	132.9	74.7	144.3	75.0
Western Europe	119.0	76.8	121.1	68.6	109.9	54.9
Absolute difference	28.9	24.3	11.8	6.1	34.4	20.1
Relative difference	124	132	110	109	131	137
Cause as proportion of difference (%)	41	23	27	8	17	10
Injury and poisoning						
Eastern Europe	108.0	43.6	106.2	41.0	117.8	45.2
Western Europe	94.3	39.3	97.6	41.2	93.4	41.1
Absolute difference	13.7	4.3	8.6	-0.2	24.4	4.1
Relative difference	115	111	109	100	126	110
Cause as proportion of difference (%)	19	4	19	0	12	2

(Notes continue on next page)

1950s and for women since the late 1960s. In Warsaw Pact Europe in 1985, moreover, standardized male and female death rates for cardiovascular disease were higher than they were in Western Europe at their peak postwar levels. A growing differential in deaths attributed to heart disease accounts by far for the greatest portion of the expanding gap between mortality levels in Communist and non-Communist Europe. These divergent trends in death from heart disease appear to explain (if only arithmetically) the discrepant paths of health change in Eastern and Western Europe over the past generation. It may be observed that the same holds true for comparisons of more specific areas (tables 4–11 and 4–12).

Potential Factors in Recent Eastern European Health Problems

How are the divergent trends in Eastern and Western European mortality—particularly in mortality attributed to cardiovascular disease—to be explained? The question might be easily answered if the proximate and underlying causes of changing patterns of heart disease among national populations were well understood. Unfortunately—and perhaps surprisingly—no such understanding can be said to exist. As WHO researchers have noted,

> A great deal of epidemiological research was initiated after the 1950s to explain the risk factors and natural history of CVD [cardiovascular disease]. However, neither these studies nor an analysis of national mortality statistics could adequately explain the dynamics of changes in CVD.[16]

As G. Lamm has pointed out, there is no "definitive answer . . . in Europe or elsewhere . . . to the question of whether decrease in [CVD]

(Continued from preceding page)

NOTES: See table 4–7 for definition of "age-standardized death rates (European Model)."

a. Eastern Europe: 1955–1959 figures for Czechoslovakia and Hungary; 1965–1969 figures for Bulgaria, Czechoslovakia, Hungary, and Poland; 1975–1979 figures for Bulgaria, Czechoslovakia, German Democratic Republic, Hungary, Poland, and Romania.

b. Western Europe: Figures for Austria, Belgium, Denmark, Federal Republic of Germany, Finland, France, Greece, Iceland, Ireland, Italy, Luxembourg, Netherlands, Norway, Portugal, Spain, Sweden, Switzerland, United Kingdom (England and Wales) with the following exceptions: 1955–1959 figures do not include Greece, Ireland, and Luxembourg; 1965–1969 and 1975–1979 figures do not include Italy.

SOURCE: For all of table 4–10: derived from World Health Organization, *World Health Statistics Annual 1988* (Geneva: WHO, 1989).

TABLE 4–11
AGE-STANDARDIZED DEATH RATES FOR SELECTED CAUSES IN THE GERMAN DEMOCRATIC REPUBLIC AND THE FEDERAL REPUBLIC OF GERMANY, 1975–1979

	Death Rate per 100,000	
Cause of Death	Male	Female
All causes		
GDR	1,426.6	956.5
FRG	1,346.6	828.4
Absolute difference (GDR-FRG)	80.0	128.1
Relative difference (FRG = 100)	106	115
Circulatory system		
GDR	741.0	551.8
FRG	608.8	394.9
Absolute difference	132.2	156.9
Relative difference	122	140
Cause as proportion of difference (%)	165	122
Respiratory system		
GDR	127.1	42.3
FRG	101.6	39.2
Absolute difference	25.5	3.1
Relative difference	125	108
Cause as proportion of difference (%)	32	3

NOTE: Earlier data and injury and poisoning data for the GDR are not available.
SOURCE: Same as table 4–10.

mortality is attributable to better treatment or to progress in prevention."[17] Indeed one may get a sense of just how little is actually understood (and agreed upon) about factors relating to heart disease by reviewing the stated primary objective of the WHO's Project for Monitoring Trends and Determinants in Cardiovascular Disease (the MONICA project, officially begun in 1984):

> To measure trends and determinants in [CVD] . . . and to assess the extent to which these trends are related to changes in known risk factors, daily living habits, health care, or major socioeconomic features measured at the same time in defined communities in different countries.[18]

TABLE 4–12
AGE-STANDARDIZED DEATH RATES FOR SELECTED CAUSES IN YUGOSLAVIA AND SELECTED SOUTHERN EUROPEAN COUNTRIES (GREECE, PORTUGAL, SPAIN), 1965–69 TO 1975–79 (EUROPEAN MODEL)
(per 100,000)

	1965–69		*1975–79*	
Cause of Death	Male	Female	Male	Female
All causes				
Yugoslavia	1,407.5	1,081.2	1,346.9	958.3
Southern European	1,333.3	955.2	1,246.8	816.3
Absolute difference (Yug.-S. Eur.)	74.2	126.0	100.1	142.0
Relative difference (S. Eur. = 100)	106	113	108	117
Circulatory system				
Yugoslavia	463.1	398.6	594.0	486.6
Southern European	460.1	372.4	501.4	367.9
Absolute difference	3.0	26.2	92.6	118.7
Relative difference	101	107	118	132
Cause as proportion of difference (%)	4	21	93	84
Respiratory system				
Yugoslavia	87.6	60.4	86.2	52.8
Southern European	143.4	91.8	123.1	67.7
Absolute difference	–55.8	–31.4	–36.9	–14.9
Relative difference	61	66	70	78
Cause as proportion of difference (%)	–75	–25	–37	–10
Injury and poisoning				
Yugoslavia	96.5	31.4	105.2	35.4
Southern European	76.6	27.3	85.7	30.4
Absolute difference	19.9	4.1	19.5	5.0
Relative difference	126	115	123	116
Cause as proportion of difference (%)	27	3	19	4

SOURCE: Same as table 4–10.

While considerable dispute and uncertainly remain over the precise etiology of cardiovascular disease in national populations (and over the correspondence between the disease and subsequent mortality), it may do well to review some of the ecological relationships in Eastern Europe between major risk factors commonly associated with cardiovascular disease and local populations. Such factors may also have a more general relevance to health conditions in Eastern Europe. Indeed, as V. Grabauskas has recently written, "a number of characteristics traditionally considered as cardiovascular risk factors have in fact a much broader negative impact on health."[19]

Smoking. Medical research and epidemiological studies have long associated heavy tobacco use with a variety of health problems.[20] One of these is increased risk of cardiovascular disease.[21] Samuel Preston has argued that the slowdown in health progress (and increase in mortality from heart disease) among older men in the United States and some other Western countries in the generation following the Great Depression can be explained (in a statistical sense) by the corresponding rise during those decades in cigarette smoking.[22]

Table 4–13 presents U.S. Department of Agriculture (USDA) estimates of annual per capita cigarette consumption for the population fifteen years of age and older for the countries of Eastern Europe. Between 1965 and 1980 per capita consumption in Warsaw Pact Europe is estimated to have risen more than a third; in Yugoslavia it is estimated to have risen more than 70 percent. Cigarette consumption in Warsaw Pact Europe, by these estimates, in 1985 was significantly higher than in a representative sample of six Western European countries; as recently as 1970 it was higher in the latter.

Available data suggest that a greater fraction of the adult population smokes in Eastern Europe than in a number of Western countries. A 1985 survey in Poland, for example, estimated that 71 percent of men and 56 percent of women aged thirty to thirty-four were smokers.[23] That same year East Germany's Committee for Health and Nutrition announced that nearly 60 percent of boys and 50 percent of girls aged fourteen to eighteen were smokers.[24] In the United States, by contrast, the 1985 figures for persons twenty-five to forty-four years of age were 42 percent for men and 31 percent for women.[25]

Eastern European populations are not only smoking more than the Western European public, but there is reason to believe that they are also smoking stronger cigarettes. In the United States the rating for tar of the sales-weighted average cigarette declined nearly two-thirds between 1954 and 1980, and the rating for nicotine by more than half;[26] much of the decline can be ascribed to the spread of the filter-tipped

TABLE 4–13
ESTIMATED ANNUAL CIGARETTE CONSUMPTION PER PERSON AGED FIFTEEN OR OLDER IN EASTERN AND WESTERN EUROPEAN COUNTRIES, 1965–1987

Country	*1965*	*1970*	*1975*	*1980*	*1985*	*1987*[a]
Bulgaria	1,431	1,494	1,944	1,855	2,366	2,225
Czechoslovakia	1,827	1,853	2,024	2,059	2,350	2,295
GDR	1,473	1,574	2,039	2,291	2,397	2,350
Hungary	2,371	2,745	3,070	3,388	3,198	3,160
Poland	2,458	2,899	3,245	3,489	3,294	3,548
Romania	1,641	1,723	1,889	2,079	2,085	1,993
Unweighted average, Eastern Europe	1,867	2,048	2,369	2,527	2,615	2,595
Yugoslavia	1,911	2,292	2,556	3,251	3,115	3,155
Unweighted average, Western Europe[b]	1,813	2,172	2,324	2,357	2,292	2,239

a. Preliminary data.
b. Western Europe: Federal Republic of Germany, France, Greece, Italy, Norway, United Kingdom.
SOURCE: U.S. Department of Agriculture Databank, 1988.

cigarette. Similar patterns can be seen in Western Europe. By contrast, as of 1980, fewer than half of the cigarettes sold in Poland were filtered; in Sweden the corresponding proportion was more than 90 percent.[27]

For Western Europe as a whole, per capita cigarette use is estimated to have been declining since the mid-1970s; in some European societies it is estimated to have been declining since the 1960s. In Warsaw Pact Europe and Yugoslavia per capita cigarette consumption stabilized in the 1980s. It is not clear, however, whether this interruption of earlier upward trends signifies a change in popular attitudes and preferences (as in Western Europe's consumption declines) or merely reflects the economic problems characteristic of the region over the past decade.

Drinking. While many of the particulars in the correspondence between drinking and heart disease[28] (and even of the correspondence between drinking and liver cirrhosis)[29] within national populations are still debated, few health specialists would contest the proposition that patterns of heavy and habitual drinking constitute a significant health risk in ordinary populations. Health problems attendant to drinking

TABLE 4–14
ESTIMATED PER CAPITA ANNUAL CONSUMPTION OF DISTILLED SPIRITS IN EASTERN EUROPE, USSR, AND WESTERN EUROPE, 1960–1980
(liters of pure alcohol)

Country	*1960*	*1970*	*1980*
Bulgaria	0.8	1.9	2.0
Czechoslovakia	1.1	2.5	3.5
GDR	1.4	2.5	4.3
Hungary	1.4	2.8	4.3
Poland	2.4	3.1	5.9
Romania	1.1	2.4	2.2
Unweighted average, Eastern Euope	1.4	2.5	3.7
Index (1960 = 100)	100	183	271[a]
USSR	4.7	6.2	6.8
Index (1960 = 100)	100	131	144
Unweighted average, 9 NATO European countries	1.2	1.8	2.2
Index (1960 = 100)	100	153	187

a. 1979.
SOURCES: M. Harvey Brenner, "International Trends in Alcohol Consumption and Related Pathologies," National Institute on Alcohol and Alcoholism, ed., *Alcohol and Health Monograph No. 1* (Washington, D.C.: U.S. Department of Health and Human Services, 1981); Werner Lelbach, "Continental Europe," in Pauline Hall, ed., *Alcoholic Liver Disease: Pathology, Epidemiology and Clinical Aspects* (New York: John Wiley and Sons, 1985); Vladimir Treml, *Alcohol in the USSR: A Statistical Study* (Durham, N.C.: Duke Press Policy Studies, 1982).

seem to be most strongly associated with the heavy and regular consumption of hard liquor.[30]

Table 4–14 presents estimates of trends in per capita consumption of distilled spirits in Eastern Europe, the USSR, and Western Europe. Around 1960, per capita consumption of hard liquor was already estimated to be higher in Warsaw Pact Europe than in Western Europe. By 1980, however, it was estimated to be dramatically higher. Between 1960 and 1980, in fact, Soviet Bloc Europe's pattern of hard spirit use seems to have edged steadily closer to a Soviet norm (table 4–14). Hard liquor, moreover, in the 1980s was the alcohol of choice in much of Warsaw Pact Europe, as in the USSR (table 4–15).

With the possible exception of Romania (where as recently as 1985 officials insisted that "alcohol consumption is not considered to be

TABLE 4–15
EUROPEAN COUNTRIES WITH DISTILLED SPIRITS AS MORE THAN ONE-THIRD OF TOTAL CONSUMPTION OF ALCOHOL, 1980
(percent)

Country	*Consumption*
Poland	69.0
USSR[a]	59.7
Iceland	57.7
Sweden	48.2
Finland	43.6
German Democratic Republic	46.4
Hungary	39.1
Czechoslovakia	36.4

a. 1979.
SOURCES: Europe: Werner Lelbach, "Continental Europe," in *Alcoholic Liver Disease: Pathology, Epidemiology and Clinical Aspects* (New York: John Wiley and Sons, 1985). USSR: Vladimir Treml, *Alcohol in the USSR: A Statistical Study* (Durham, N.C.: Duke Press Policy Studies, 1982).

giving rise to serious health, social, or economic problems,")[31] authorities throughout Soviet Bloc Europe have been voicing growing concern about the alcohol habits of the populations beneath them. In expressing their concern, they have also provided details about the scope of the problem. A few illustrative examples may suffice.

In Bulgaria the incidence of cirrhosis of the liver and related diseases is officially reported to have risen by an order of magnitude between 1974 and 1975. Four times as many women were said to be drinkers in the mid-1980s as in the mid-1970s. A survey in one region by Bulgaria's Communist Youth League concluded that almost 80 percent of the eighteen- to thirty-year-olds included were regular drinkers. And while such language does not lend itself to precise calibrations, one Bulgarian health official recently told a Western reporter that every third person in the capital city of Sofia has a drinking problem.[32]

In Czechoslovakia, according to Radio Prague, about 30 to 40 percent of the adult male population in industrial areas in 1986 were said to drink "excessively." The previous year, a leading Czech paper lamented that "it has long since stopped being true that only men drink."[33]

In East Germany the official medical journal *Deine Gesundheit* declared in 1987 that alcohol consumption in the country had assumed "alarming proportions." As of the mid-1980s, according to another publication, the life expectancy of the country's alcoholics was "ten to twelve years less than the average." According to an internal 1983 report of the East German Ministry of Health that was subsequently published in West Germany, one person in twelve in the GDR was deemed to be a heavy drinker, a third of these being termed untreatable alcoholics.[34]

In Hungary, according to a discussion in the National Assembly in 1986, about half a million persons were alcoholics: more than 6 percent of the population fifteen years of age or older. In 1987 a study published in the magazine of Hungary's Communist Youth League reported that among people aged thirty-one to forty about a fifth of all women and almost three-quarters of all men were heavy drinkers. (By way of comparison, a 1985 study rated 19 percent of men and 9 percent of women aged thirty-one to forty as "heavier drinkers" in the United States.)[35] Feminization of alcoholism seems to have proceeded apace; Hungary's mortality statistics, which are perhaps the most detailed and reliable in Eastern Europe, place the standardized death rate for cirrhosis at a higher level for women in the early 1980s than for men in the mid-1960s.[36] Recently Hungarian authorities have reported some successes in controlling public drunkenness: in 1987 a spot breathalyzer check of 17,000 workers showed that only 2.2 percent of them were drunk on the job. As recently as 1985 the corresponding figure had been more than 9 percent.[37]

In Poland the government estimates that about one million persons are "regular alcoholics": this would be about 4 percent of the population fifteen years of age or older. For a variety of reasons household budget surveys in centrally planned economies are problematic under even the best of circumstances; nonetheless Warsaw's Main Office of Statistics reported that the portion of total personal consumption expenditures allocated to alcohol exceeded 17 percent in 1983.[38] Although the Communist government in Poland attempted to curtail alcohol consumption after its dissolution of the Solidarity Union and its declaration of martial law in 1981, its measures did not appear to have been tremendously effective. One reason may be that the Polish state had a real financial interest in heavy drinking among the local population. Like all Warsaw Pact governments, Poland derived a considerable portion of its state revenues from the sale of alcoholic beverages through the state liquor monopoly. In 1985, such sales accounted for 18 percent of total state revenue in Poland.[39]

Health Care

Properly framed and implemented, national health policies can control and reduce mortality levels for national populations, even during periods of increased health risks (be these self-inflicted or otherwise). Eastern European health policies, however, have not been adequate to this task. The secular rise in mortality levels for Eastern Europe's adult population over the past two decades attests precisely to the dimensions of health policy failure in the region.

A striking pattern has emerged in recent years for Eastern European health services. The number of medical personnel per 10,000 local population has risen rapidly in the region in recent decades—indeed a good deal more rapidly than in Western Europe. At the same time general mortality levels for Eastern European adults have been rising. On the basis of epidemiological reasoning alone, one might well be led to the conclusion that Eastern Europe's doctors are hazardous to the health (table 4–16).

The negative correlation between availability of medical personnel and adult mortality levels in Eastern Europe over the past generation may speak to underlying problems in the medical and health strategies embraced and pursued by local regimes. In varying degree, Eastern Europe's medical systems were replicas of the Soviet original. The Soviet health system was originally established to deal with the health problems of a population with a life expectancy roughly the same as Ethiopia's today. It emphasized mass campaigns to control communicable and infectious diseases and made extensive use of personnel with only brief exposure to medical or public health training. The pattern of disease in modern industrial societies, however, differs dramatically from that encountered by Soviet revolutionaries when they were establishing their socialized health service. More intensive training and vastly more expensive procedures and equipment are typically required to treat the diseases that may be expected to afflict a population where life expectancy at birth approaches or exceeds seventy years. Soviet-style health systems, unfortunately, have not fully adjusted to this reality. On the contrary rising mortality in Eastern Europe attests in some measure to the mismatch of the labor-extensive, low-cost approach of the Soviet health model and the actual needs of the local populations.

One may wonder about the reasons that Soviet-style health systems have failed to respond more effectively to the health problems of the populations they are charged with serving. One factor may be ideological. Soviet doctrine assigns health care and related services to the "nonproductive sphere" of the economy. In times of economic austerity or budgetary stress, there may be pressure to reduce allocations to these supposedly nonproductive services. According to official data from the Council for Mutual Economic Assistance (CMEA or Comecon), the proportion of public consumption funds allocated to free public

TABLE 4–16
MEDICAL PERSONNEL IN EASTERN AND WESTERN EUROPE, 1960–1985
(per 10,000 population)

Country	*1960*	*1970*	*1980*	*1985*
Bulgaria	17.0	22.2	30.0	35.1
Czechoslovakia	17.5	22.2	30.0	36.0
GDR	12.1	20.3	26.1	29.9
Hungary	15.3	22.1	28.1	31.5
Poland	12.7	19.3	22.5	24.3
Romania	13.5	14.7	17.9	20.8
Unweighted average, Eastern Europe[a]	14.7	20.3	26.2	29.6
Unweighted average, Western Europe[b]	11.8	13.8	14.4[c]	NA
Ratio (Western Europe = 100)	125	147	182	NA
Yugoslavia	6.2	10.0	14.7	NA
Unweighted average, Greece, Portugal, Spain	10.8	12.8	24.3[c]	NA
Ratio (Greece,Portugal, Spain = 100)	57	78	60	NA

NA = not available.
a. Figures for Eastern Europe include doctors and dentists; figures for Western Europe exclude dentists.
b. Western Europe: Austria, Belgium, Denmark, Federal Republic of Germany, Finland, France, Greece, Iceland, Ireland, Italy, Luxembourg, Netherlands, Norway, Portugal, Spain, Sweden, Switzerland, United Kingdom.
c. Figure is for 1981 and does not include Iceland or Luxembourg.
SOURCES: Eastern Europe: Council for Mutual Economic Assistance Secretariat, *Statisticheckii Ezhegodnik Stran-Chlenov Sovieta Ekonomicheskoi Vzaimopomoshchi 1987* (Moscow: Finansy i Statistika, 1987). Western Europe, Yugoslavia: World Bank, *World Tables*, vol. 2, 3rd ed. (Baltimore: Johns Hopkins University Press, 1983).

health care and related services actually fell in Warsaw Pact Europe between 1965 and 1985 (table 4–17). While translating these allocations into a Western-style market-economy framework may present the analyst with insuperable conceptual problems, attempts to do so have nevertheless been made. One such effort is presented in table 4–18. It suggests a steadily growing gap between Warsaw Pact Europe and Western Europe in relative allocation of national resources to health care over the past two decades.

Eastern Europe's labor-extensive health strategy, it seems, may

TABLE 4–17
RESOURCE ALLOCATION TO HEALTH SECTOR BY VARIOUS MEASURES IN EASTERN AND WESTERN EUROPE, 1965–1985

Country	*1965*	*1970*	*1975*	*1980*	*1985*
A. Official Estimates of Percentage of Public Consumption Funds Allocated to Free Public Health and Physical Education					
Bulgaria	14.1	13.4	14.4	16.3	16.3
Czechoslovakia	14.7	15.0	15.2	15.7	16.0
GDR	17.7	15.3	15.8	17.9	18.8
Hungary	22.9	16.7	14.5	13.8	14.0
Poland	25.1	25.5	26.9	25.7	21.7
Romania	NA	NA	NA	NA	NA
Unweighted average, Eastern Europe	18.9	17.2	17.4	17.9	17.4
B. Estimates of Health Sector Expenditures as Percentage of National Output (Western National Income Framework)					
Unweighted average, Eastern Europe[a]	3.2	3.1	3.1	3.3	3.7
Unweighted average, Western Europe[b]	4.5	5.5	6.5	7.0	7.3[c]

a. Percentage of estimated GNP.
b. Percentage of GDP. Western Europe: Austria, Belgium, Denmark, Federal Republic of Germany, Finland, France, Greece, Iceland, Ireland, Italy, Luxembourg, Netherlands, Norway, Sweden, Switzerland, United Kingdom. 1965 figure does not include Luxembourg, Portugal; 1970 figure does not include Portugal; 1975 figure does not include Iceland; 1983 figure does not include Iceland, Ireland, Luxembourg, Portugal, Spain, Switzerland.
c. 1983 figure.
SOURCES: Eastern Europe, section A: Council for Mutual Economic Assistance Secretariat, *Statisticheskii Ezhegodnik Stran-Chlenov Sovieta Ekonomicheskoi Vzaimopomoshchi 1984* and *1987* (Moscow: Finansy i Statistika, 1984 and 1987, respectively). Eastern Europe, section B: Research Project on National Income in East Central Europe, *Eastern Europe: Domestic Final Uses of Gross Product, 1970 and 1975–1985*, Occasional Paper 92 (New York: L. W. International Financial Research, 1986). Western Europe, sections A, B: OECD, *Measuring Health Care 1960–1983* (Paris: OECD, 1985); World Bank, *World Development Report 1987* (Washington, D.C.: World Bank, 1987).

not be a complement to an upgrading of health care among state budgetary priorities but rather a substitute for it. In its Eastern European variant, socialized medicine seems to cut two ways for its patients. As Sophia Miskiewicz has noted, "[It] is financed directly and almost entirely by the state; as a result, the quantity and quality of these

TABLE 4–18
MORTALITY CHANGES AND OFFICIAL MEASURES OF ECONOMIC CHANGE IN EASTERN AND WESTERN EUROPE, 1955–1985
(percent)

Area	*1955/59–1965/69*	*1965/69–1975/79*	*1975/79–1985*
Warsaw Pact Europe			
Age-standardized mortality rate (European model)	–9	+2	+3
Per capita net material product produced (unweighted average)[a]	+73	+92	+30
OECD Europe			
Age-standardized mortality rate (European model)	–8	–11	–11
Per capita GNP (weighted average)	+44	+33	+11

NOTES: Age-standardized mortality rates are unweighted arithmetic averages for male and female rates.
a. Figures are for 1955–1965, 1965–1975, and 1975–1985, respectively.
SOURCES: Data derived from various issues of the following series: World Health Organization, *World Health Statistics Annual* (Geneva: WHO); Organization for Economic Cooperation and Development, *National Accounts* (Paris: OECD); Council for Mutual Economic Assistance Secretariat, *Statisticheskii Ezhegodnik Stran-Chlenov Sovieta Ekonomicheskoi Vzaimopomoshchi* (Moscow: Finansy i Statistika).

services are determined by the authorities in the light of political priorities."[40]

Conclusions

To a lesser but nevertheless unmistakable degree, Eastern Europe was beset by the syndrome of deteriorating conditions of public health that is now officially recognized as having afflicted the USSR. Declining life expectancy at birth, rising levels of adult mortality, and (as some recent Soviet pronouncements suggest)[41] an increase of cardiovascular mortality among adults were in 1985 common to Yugoslavia, Warsaw Pact Europe, and the USSR—and among contemporary industrialized societies, unique to them.

Epidemiological reasoning would prompt profound questions about the impact of governance on health conditions in these areas. The populations affected by rising age-adjusted mortality, after all, had different languages, cultures, and histories; the societies in question varied in material and technical attainment. The most obvious common characteristic of these countries was that they were all ruled at the time by Marxist-Leninist states, and by that particular variant of Marxist-Leninist state that came to power with the direct assistance of the Red Army. After more than two decades of health decline for adult populations in the region, it is perhaps not premature to inquire whether the health problems evidenced in these countries might have been in part systemic. Does something intrinsic to what historical materialists might term "the mature stage of socialism" have an adverse impact on the health of local populations?

Ordinarily some correspondence between stated economic progress and health progress in a region may reasonably be expected. For Europe as a whole, however, such a correspondence appears to have broken down in the 1960s. In Western Europe, reduction of mortality has proceeded with economic growth; indeed, mortality decline has accelerated even as the pace of measured per capita growth has slowed. In Eastern Europe, however, age-adjusted mortality rates had been rising over a period in which substantial increments in economic output have been officially claimed. This dissonance might prompt reassessment of the actual significance of the achievements of Eastern European regimes that are currently recorded in official statistical yearbooks. If such a reassessment were to conclude that material progress did indeed occur between the mid-1960s and the mid-1980s, that in itself would prompt a subsidiary set of questions about the relation between socioeconomic performance and health in Eastern Europe.

In Western countries medical research today is increasingly concerned with the impact of psychological and emotional factors in health and disease. The psychosocial aspect of cardiovascular disease, for example, is a topic of serious and active interest on the part of U.S. health authorities.[42] A number of studies seem to suggest, at least to some researchers, that such intangible factors as attitude, outlook, and satisfaction with life may play a more important role in physical well-being than was previously believed. A. Jablensky argues that "there are good reasons to surmise that, owing to methodological difficulties, results reported up to date may be, in fact, an under-estimate of the actual contribution of psychological factors to cardiovascular morbidity."[43] What further research and improved methodologies will reveal remains to be seen. Such findings, however, may prove to be of particular interest and significance to the apparently increasingly unhealthy populations of formerly Communist Europe.

5

Demographic Factors in Soviet Power

Soviet strategists typically analyze and discuss the USSR's prospects in its extended international struggle in terms of what they call the correlation of forces. At once more sweeping and more intricate than the notion of a balance of power, this concept suggests assessment of the power of states on the basis of material, political, and even psychological factors. Implicit in this concept is the view that trends in the international arena are affected by the respective abilities of states engaged in political combat to mobilize and to apply strength drawn from their domestic base.[1]

From the standpoint of the correlation of forces, demographic trends in the USSR's domestic base do not augur favorably for the Soviet state. A conjuncture of several different trends has unexpectedly complicated the talk of augmenting deployable power. While some of these problems may be transient, others as yet show no signs of resolution. Moreover, since demographic trends involve changes that are by their nature gradual and long term, most of these current problems may be expected to affect Soviet power continuously, well into the next century—indeed throughout the life course of today's Soviet population, whose demographic contours have already been touched by the rhythms and problems in question. Only three types of such internal demographic trends that may bear upon the correlation of forces are examined here: fertility, mortality, and labor force.

This chapter, previously unpublished, was written in 1987. It is reprinted as originally presented to the International Security Council conference, Geneva, September 1987.

TABLE 5–1
POPULATION GROWTH, 1950–1985, AND DOUBLING TIMES AT 1979–1985 GROWTH RATES IN THE USSR AND ITS REPUBLICS
(population in thousands)

Area	*1985 Population*	*% of Total Population*	*1950–80 Annual Increase (%)*	*1979–85 Annual Increase (%)*	*Doubling Time (Years)*
USSR	278,855	100.0	1.3	0.9	77
Slavic republics		(73.6)			
RSFSR	143,860	51.6	1.0	0.7	103
Ukraine	51,184	18.4	1.0	0.4	176
Belorussia	10,027	3.6	0.7	0.7	98
Moldavia	4,170	1.5	1.8	0.8	84
Baltic republics		(2.8)			
Estonia	1,537	0.6	1.0	0.7	101
Latvia	2,610	0.9	0.9	0.5	143
Lithuania	3,587	1.3	0.9	0.8	89
Transcaucasus		(5.5)			
Armenia	3,365	1.2	2.8	1.6	44
Azerbaydzhan	6,750	2.4	2.5	1.7	41
Georgia	5,270	1.9	1.2	0.7	95
Kazakhstan	16,126	5.8	2.7	1.4	49
Central Asia		(11.0)			
Kirgiziya	4,061	1.5	2.5	2.1	33
Tadzhikistan	4,620	1.7	3.2	3.0	24
Turkmenistan	3,280	1.2	2.9	2.6	27
Uzbekistan	18,410	6.6	3.1	2.7	26

SOURCES: 1950–1980 estimates from Godfrey S. Baldwin and Stephen Rapawy, "Demographic Trends in the Soviet Union," in U.S. Congress, Joint Economic Committee, *Soviet Economy in the 1980s: Problems and Prospects,* 97th Congress, 2nd session, December 31, 1982, vol. 2, p. 267. 1979–1985 estimates from W. Ward Kingkade, "Estimates and Projections of the Population of the USSR by Age and Sex for Union Republics, 1970 to 2025" (U.S. Census Bureau, 1987, unpublished paper).

Fertility

Table 5–1 summarizes current population levels and rates of increase for the USSR's component socialist republics. Like other countries with an advanced industrial base, the Soviet Union's rate of annual population increase has tended to drop since the end of the early postwar period, and its doubling time, like theirs, has correspondingly risen. With a current doubling time of about seventy-seven years, the USSR's pace of population growth is similar to that of the United States and of several other Western countries.[2]

This aggregate rate of population growth, however, disguises enormous variations between republics. In the Slavic and Baltic republics, where populations are overwhelmingly of European ethnic descent, current doubling times are generally close to a century, in several cases considerably slower. By contrast, in Central Asia, where the overwhelming majority of population is of Islamic heritage, most doubling times are less than thirty years. (In Kazakhstan, a traditionally Muslim area where Soviet policy directed a massive influx of European nationals, doubling time is between these extremes, representing an average of two distinct patterns.) The difference in population growth rates speaks primarily to the differences in fertility among the USSR's disparate ethnic groups.

Tables 5–2 and 5–3 illustrate these differences. In the decades since recovery from World War II, general fertility for the Soviet Union has fallen from about three children per woman to about two and a half children per woman—a level still above net replacement. In the European republics, however, fertility rates appear to have fallen to replacement, or below, in the 1970s and early 1980s. In the Central Asian republics, conversely, the total fertility rate (children born per woman of childbearing ages) was well over five for much of the postwar period. It appears to have risen substantially in these areas during the 1960s, perhaps in response to improvements in living standards, although improvements in enumeration may also account for much of this apparent rise. (Birth rates tell a similar story.)

Fertility rates for the republics understate differences in fertility between the Soviet Union's ethnic groups. The figures for the Russian republic, for example, are raised by the presence of a large number of Muslim inhabitants; as many as a quarter of the Soviet Union's almost 50 million "Muslims" live in the Russian Soviet Federated Socialist Republic (RSFSR).[3] In Central Asia, total averages are depressed by the European presence; fertility is lowest among the four in Tadzhikistan because its fraction of inhabitants of Russian nationality is the highest. Thus fertility for the European population may be distinctly below replacement, while the total fertility rate for Islamic ethnic groups may stand at well over five children per woman—a higher level than the one the World Bank currently imputes to India.[4]

Demographers commonly expect prosperity and improved living

TABLE 5–2
FERTILITY RATES IN THE USSR AND ITS REPUBLICS, 1958–1983

USSR and Republics	*1958–59*	*1969–70*	*1979–80*	*1983*
USSR	2.8	2.5	2.3	2.5
Slavic republics				
RSFSR	2.6	2.0	1.9	2.1
Ukraine	2.3	2.0	2.0	2.1
Belorussia	2.8	2.3	2.0	2.1
Moldavia	3.6	2.6	2.4	2.6
Baltic republics				
Estonia	1.9	2.1	2.0	2.2
Latvia	1.9	1.9	1.9	2.2
Lithuania	2.6	2.4	2.0	2.0
Transcaucasus				
Armenia	4.7	3.2	2.4	2.4
Azerbaydzhan	5.0	4.6	3.3	3.1
Georgia	2.6	2.6	2.2	2.2
Kazakhstan	3.3	3.3	2.9	2.9
Central Asia				
Kirgiziya	4.3	4.8	4.1	4.1
Tadzhikistan	3.9	5.9	5.8	5.5
Turkmenistan	5.1	5.9	5.1	4.9
Uzbekistan	5.0	5.7	4.9	4.7

NOTE: The rate is for children per woman aged 15–49. Rates are two-year averages, except for 1983.
SOURCES: Estimates for 1958–1980 from Murray Feshbach, "The Soviet Union: Population Trends and Dilemmas," *Population Bulletin*, vol. 37, no. 3, 1982, pp. 20–21. Estimates for 1983 from W. Ward Kingkade, "Estimates and Projections of the Population of the USSR by Age and Sex for Union Republics, 1970 to 2025" (U.S. Census Bureau, 1987, unpublished paper), table 1.

standards to reduce fertility among populations where fertility levels are high. (This in essence is the theory of the demographic transition.) To date, the USSR's Muslim populations have proved an exception to this rule; despite improvements in health, education, nutrition, housing, and income, their fertility has remained close to traditional levels. Some Western analysts believe there are today signs of an incipient fertility decline among the USSR's Islamic peoples.[5] Even if one posits a steady decline in their fertility over coming decades, however, the

TABLE 5–3

CRUDE BIRTH RATES IN THE USSR AND SELECTED REPUBLICS, 1950–1980
(births per 1,000 population)

USSR and Republics	*1950*	*1960*	*1970*	*1980*
USSR	26.7	24.9	17.4	18.3
Slavic republics				
RSFSR	26.9	23.2	14.6	15.9
Ukraine	22.8	20.5	15.2	14.8
Belorussia	25.5	24.4	16.2	16.0
Kazakhstan	37.6	37.2	23.4	23.8
Central Asia				
Kirgiziya	32.4	36.9	30.5	29.6
Tadzhikistan	30.4	33.5	34.8	37.0
Turkmenistan	38.2	42.4	35.2	34.3
Uzbekistan	30.8	39.8	33.6	33.8

SOURCE: Murray Feshbach, "The Soviet Union: Population Trends and Dilemmas," *Population Bulletin*, vol. 37, no. 3 (1982), p. 17.

remaining differentials will have a profound impact on the composition of the Soviet population.

This may be seen in table 5–4. These U.S. Census Bureau projections assume a gradual decline in fertility for Soviet Central Asia (with zero population growth in 2050). Nevertheless changes in population composition are dramatic. In 1985 about 52 percent of the USSR's inhabitants lived in its Russian republic. But that republic accounted for only about 46 percent of the Soviet Union's children under age five—and some of those children, as has been noted, were sons and daughters of parents of Muslim background. By contrast, Central Asia, with 11 percent of the 1985 population, accounted for 19 percent of the USSR's young children; adding in Kazakhstan produces a figure of 26 percent. Projecting to the year 2000, the Russian republic will likely have only about 40 percent of the USSR's young children, while the Muslim republics and Kazakhstan will account for about 33 percent. For the year 2025 the same series of projections gives a figure of 36 percent for the RSFSR and of 39 percent for the "Muslim" republics and Kazakhstan.[6]

At present the Russian ethnic group may just barely represent a majority of the population in the USSR (although it may already constitute less than half of the total Soviet population). This fraction can be expected to diminish progressively. Barring catastrophe (a

TABLE 5-4
PROJECTED POPULATION UNDER FIVE YEARS OF AGE IN THE USSR AND SELECTED REPUBLICS AND REGIONS, 1985–2000
(thousands)

Area	*1985*	*1990*	*1995*	*2000*
USSR	26,032	26,410	24,498	24,521
RSFSR	11,884	11,409	9,921	9,707
Kazakhstan	1,854	1,935	1,904	1,938
Central Asia	4,864	5,494	5,791	6,190
RSFSR population as % of USSR population	45.7	43.2	40.5	39.6
Central Asia and Kazakhstan population as % of USSR population	25.8	28.1	31.4	33.1

SOURCE: W. Ward Kingkade, "Estimates and Projections of the Population of the USSR by Age and Sex for Union Republics, 1970 to 2025" (U.S. Census Bureau, 1987, unpublished paper).

circumstance admittedly not unknown in Soviet history), Russian babies may be outnumbered by "Muslim" babies in the second decade of the coming century.

Some observers have taken these trends to mean that increasing unrest among the USSR's "Muslims" is likely, and even that an internal revolt may be in the making. These observers may ultimately prove to be correct, although there would seem to be little in postwar Soviet history to lend support to such an apocalyptic interpretation.

Expected changes in the population composition of the USSR, however, would present Soviet administrators with significant problems even under conditions of complete political tranquility. A multiethnic Red Army with a diminishing Russian presence would face increasingly complex logistical issues; the question of communications could become all the more salient in light of the enduring problems with Russian-language training in the Soviet Central Asian educational system.[7]

Contrary to the popular impression in the West, the USSR's "Muslim" areas receive substantial subsidies and transfers[8]—in effect from the other regions of the USSR (principally from the Russian republic). Maintaining these subsidies would become more onerous over time. Yet reducing them might seem unwise, at least to the extent that they are viewed locally as part of a bargain that purchases social peace.

Finally, with a changing population composition, the quality, nature, and location of the USSR's labor force will also be changing

TABLE 5–5
CHANGES IN DEATH RATES AT SPECIFIC AGES IN THE USSR, 1964–1985
(per 1,000 population)

Age	*1964–65*	*1984–85*	*Change (%)*
20–24	1.6	1.5	–3
25–29	2.0	2.0	+3
30–34	2.6	2.8	+10
35–39	3.3	3.6	+11
40–44	4.1	5.7	+39
45–49	5.5	7.3	+33
50–54	8.7	11.3	+31
55–59	12.0	15.1	+26
60–64	19.4	20.4	+5
65–69	27.5	31.1	+13
70+	71.0	78.7	+11

SOURCES: 1964–1965 data from Murray Feshbach, *A Compendium of Soviet Health Statistics,* U.S. Census Bureau, Center for International Research, Report no. 9, (1985), p. 59. 1984–1985 data from *Vestnik Statistiki* 1986, no. 12, cited in W. Ward Kingkade, "Recent Trends in Soviet Adult Mortality" (U.S. Census Bureau, 1987, unpublished paper), table 2.

in ways that will not be expected to stimulate growth of the Soviet economy.

Mortality

Like other countries with an advanced industrial base, the Soviet Union experienced a general decline in mortality in the 1950s. Improvements in health in the USSR were in fact unusually rapid during that period; according to the United Nations Population Division, life expectancy had reached, and possibly exceeded, that of the United States.[9]

In the mid-1960s, however, mortality rates by age group were beginning instead to rise. The phenomenon first became visible for cohorts of middle-aged men; the pattern spread until virtually all adult age groups were affected. When reported infant mortality rates began to rise in the early 1970s, Soviet authorities began to impose something like a statistical blackout over data pertaining to mortality. In the darkest days of this blackout (the early and mid-1980s) it became difficult to obtain even such simple figures as the country's crude death rate. (As it happens, the officially reported crude death rate rose more than 56 percent between 1964 and 1985—far too great a jump to be explained by aging of the population alone.)

TABLE 5–6
LIFE EXPECTANCIES AT BIRTH IN THE USSR, 1958–1986
(years)

Date	*Both Sexes*	*Male*	*Female*
Official figures			
1958–59	69	64	72
1971–72	70	64	74
1978–79	68	62	73
1983–84	68	63	73
1984–85	68	63	73
1985–86	69	64	73
U.S. Census Bureau estimates			
1970	68.7	64.3	73.3
1979	67.5	62.5	72.8
1985	67.4	62.6	72.5

SOURCE: W. Ward Kingkade, "Recent Trends in Soviet Adult Mortality" (U.S. Census Bureau, 1987, unpublished paper).

When age-specific death rates for the USSR were published in 1986 for the first time in virtually a decade, it was apparent that almost all adult cohorts had suffered a pronounced and sustained deterioration in health (table 5–5). These same data indicated that death rates for children under five had risen by more than 11 percent between 1969–1970 and 1984–1985,[10] suggesting that infant mortality in the USSR in the mid-1980s may have been no lower than it had been twenty years earlier.

The net impact of these rising mortality rates was to lower life expectancy in the USSR. Recently released Soviet figures place life expectancy at 69 years in 1985–1986; according to the same series, it had been 70 years in 1971–1972 (table 5–6). A recent evaluation of Soviet demographic data by an analyst at the U.S. Census Bureau, however, suggests that combined male and female life expectancy at birth in the USSR in 1985 may actually have been as low as 67.4 years—more than 7 years below the contemporary level in the United States, and lower than the life expectancy numbers the World Bank currently gives such places as Chile, Jamaica, and Trinidad and Tobago.[11] If these Census Bureau estimates are accurate, a boy born today would have a lower life expectancy in the USSR than in Mexico.

By Census Bureau estimates, life expectancy in the Soviet Union fell four years for men and about two years for women between the mid-1960s and the mid-1980s. No country with an advanced industrial

TABLE 5–7

RATIO OF INCIDENCE OF INFECTIOUS DISEASES, THE USSR TO THE UNITED STATES, 1970 AND 1979

(per 100,000 population)

Disease	*1970*	*1979*
Typhoid fever	52.9:1	29.2:1
Diphtheria	2.1:1	3.3:1
Whooping cough	7.7:1	13.5:1
Tetanus	3.9:1	3.8:1
Polio (acute: paralytic)	5.5:1	8.0:1
Measles	8.4:1	23.4:1

SOURCE: Murray Feshbach, "Issues in Soviet Health Problems," in U.S. Congress, Joint Economic Committee, *Soviet Economy in the 1980s: Problems and Prospects*, 97th Congress, 2d session, December 31, 1982, vol. 2, p. 223.

base had ever before suffered such a peacetime deterioration in public health.

The causes of the Soviet Union's broad-based rises in mortality have been widely discussed and debated in the West.[12] Increases in alcohol consumption, tobacco, and drug use; environmental deterioration; and rises in work-related accidents are but some of the possible explanations. Whatever the relative effect of these factors, a pervasive and long-term deterioration in public health could not take place if the country's medical system were not also beset by serious difficulties. As table 5–7 indicates, the prevalence of infectious and easily controlled diseases was markedly higher in the USSR than in the United States in 1979; many of these disparities have subsequently increased. Evidence suggests that the failure of the USSR's socialized medical sector to check increases in national mortality speaks not only to administrative issues but also to questions of basic strategy. The USSR actually seems to be devoting a smaller share of its national resources to health care today than twenty-five years ago.[13]

If there is a general failure of effectiveness today in the Soviet medical system, that failure would nevertheless seem to be less than universal. In certain republics, such as Georgia and Armenia, estimated life expectancy is much what one would expect of a country in Western Europe (table 5–8). In Turkmenistan, conversely, U.S. Census Bureau estimates of life expectancy at birth are no higher than the figures the World Bank currently gives for such Middle Eastern countries as Turkey and Jordan and lower than World Bank figures for Lebanon and Syria.[14] Significantly, the Russian republic's life expectancy levels may be closer

TABLE 5–8
LIFE EXPECTANCIES AT BIRTH IN THE USSR AND ITS REPUBLICS, 1975 AND 1983
(years)

	1975		*1983*	
USSR and Republics	Male	Female	Male	Female
USSR	63	74	62[a]	73[a]
Slavic republics				
RSFSR	62	74	62	75
Ukraine	65	74	65	76
Belorussia	67	77	65	78
Moldavia	63	69	62	69
Baltic republics				
Estonia	65	75	64	76
Latvia	64	75	64	76
Lithuania	67	77	66	77
Transcaucasus				
Armenia	70	77	69	78
Azerbaydzhan	64	72	65	75
Georgia	68	77	68	79
Kazakhstan	63	75	62	76
Central Asia				
Kirgiziya	62	72	62	73
Tadzhikistan	62	68	62	70
Turkmenistan	62	69	60	68
Uzbekistan	64	72	63	72

a. Not directly comparable with estimates for republics.
SOURCES: 1975 estimates from Godfrey S. Baldwin, *Population Projections by Age and Sex: For the Republics and Major Economic Regions of the USSR: 1970–2000* (Washington, D.C.: U.S. Census Bureau, 1979), p. 22. 1983 republic estimates from W. Ward Kingkade, "Estimates and Projections of the Population of the USSR by Age and Sex for Union Republics, 1970 to 2025" (U.S. Census Bureau, 1987, unpublished paper), table 1. 1983 estimate for Soviet life expectancy interpolated from U.S. Census Bureau estimates for 1979 and 1985.

to Central Asian than to other European republics'.

The consequences of increased mortality are diverse. Rising age-adjusted death rates have slowed the growth of the Soviet population—thus the growth of the population of military and working age. Moreover, the rise in mortality, with the especially conspicuous decline in life expectancy for the Russian republic, may have hastened shifts in the ethnic composition of the Soviet population.

Such trends have attracted the attention of the military. In February 1983, Full Admiral Sorokin, then first deputy chief of the Political-Military Administration of the Soviet Armed Forces, wrote of "the unsatisfactory demographic situation" and of the need "to overcome the negative impact [of it] on the combat capability of the Soviet Armed Forces."[15]

As for the civilian sector, declining levels of health complicate, and perhaps compromise, attempts to accelerate economic growth. To the extent that improvements in health constitute increments in human capital, and thereby facilitate improvements in productivity, the converse may also hold true for deteriorating health conditions. The problem would seem all the more salient when declines in public health are particularly pronounced among the population of working ages.

The health decline in the USSR is surely amenable to correction. But the policies necessary to halt and to reverse it may also entail consequences and costs for Soviet administration. This may explain the reluctance to date to address the problem frontally.

Labor Force

Throughout the early postwar period, as the USSR's population of working ages (defined in the USSR as sixteen to fifty-nine for men, sixteen to fifty-four for women) grew at a fairly steady pace; for 1950–1980 its average annual rate of growth was more than 1.3 percent. In the 1980s, however, through a conjuncture of events ("echo effects" from World War II, declining fertility in the 1960s, rising mortality schedules) growth in the populations of working ages has been sharply curtailed (table 5–9). The 1980s saw a growth in population of working ages of about 0.3 percent per annum—less than a quarter of the pace of the previous thirty years. The slow pace of growth will continue until at least the late 1990s. (The prediction can be made with some confidence, insofar as the work force of the late 1990s is already alive today.)

In the past, Soviet economic growth has been extensive; that is to say, it has relied heavily upon the mobilization of growing or untapped pools of labor and upon augmentation of stocks of capital. Extensive growth is no longer possible, at least with labor. Not only is the pace of growth slowing for the population of working ages, but the rates of labor force participation for both men and women are already very high by any international perspective (tables 5–10 and 5–11). (The Soviet Union's teenage labor force participation rates may look somewhat low, but conscription for military service lowers

TABLE 5–9
WORKING-AGE POPULATION OF THE USSR AND AVERAGE ANNUAL GROWTH RATE, 1950–1985
(thousands)

Year	*Total of Working Age*	*% of Total Population*	*Males of Working Age*	*% of Total Male Population*	*Females of Working Age*	*% of Total Female Population*	*Females as % of Total Working Age*
1950	103,345	57.4	44,867	56.7	58,478	58.0	56.6
1955	114,658	58.5	51,502	58.9	63,156	58.1	55.1
1960	119,459	55.7	55,381	57.1	64,078	54.6	53.6
1965	124,142	53.8	59,169	56.0	64,973	51.8	52.3
1970	131,645	54.2	64,653	57.7	66,992	51.2	50.9
1975	144,667	56.8	71,491	60.5	73,176	53.7	50.6
1980	153,044	57.9	76,905	62.4	76,139	54.0	49.7
1985	157,295	56.4	81,322	62.1	75,973	51.4	48.3

Total working-age population, average annual increase (%)

1950–60	1.5
1960–70	1.0
1970–80	1.5
1980–85	0.3

NOTE: Officially defined as males 16–59 years of age and females 16–54 years of age, inclusive. Data as of July 1.
SOURCES: 1950–1965 estimates from Stephen Rapawy, *Estimates and Projections of the Labor Force and Civilian Employment in the USSR, 1950–1990*, Foreign Economic Report, no. 10 (U.S. Department of Commerce, 1976), p. 4. 1970–1985 estimates from W. Ward Kingkade, "Estimates and Projections of the Population of the USSR by Age and Sex for Union Republics, 1970 to 2025" (U.S. Census Bureau, 1987, unpublished paper).

this figure about twenty points.)

The composition of the cohorts rising into the labor force is distinctly different from those retiring from it, as may be seen in table 5–12. Between 1980 and the end of the century, the working age population of the Slavic republics (RSFSR, Belorussia, Ukraine) is anticipated to decline slightly. More than four-fifths of the net addition to Soviet working age population is anticipated to come from Kazakhstan and the Central Asian republics, and persons of Muslim heritage may account for at least as high a fraction of net working age population growth during this period.

From the standpoint of Soviet planning, this shift in work force

TABLE 5–10
LABOR FORCE PARTICIPATION RATES IN THE USSR, 1971 AND 1981
(percent)

	Male		*Female*	
Age	1971	1981	1971	1981
16–19	53.3	48.4	47.8	40.8
20–29	89.7	89.5	86.3	86.1
30–39	97.6	97.6	92.7	92.7
40–49	95.9	95.9	90.6	90.6
50–54	90.0	90.0	77.3	77.3
55–59	79.9	79.9	44.4	45.4
60+	49.0	50.0	25.0	26.0

SOURCE: Ann Goodman and Geoffrey Schleifer, "The Soviet Labor Market in the 1980s," in U.S. Congress, Joint Economic Committee, *Soviet Economy in the 1980s: Problems and Prospects*, 97th Congress, 2d session, December 31, 1982, vol. 2 , p. 328.

composition is not fortuitous. Educational levels are generally lower in Muslim republics than in European areas. Russian language training, as already mentioned, remains a problem. And the populations of Islamic descent are widely thought to be more reluctant to participate in the socialist sector of the economy, preferring the various shadings of private activity.[16] While such a preference would not necessarily

TABLE 5–11
LABOR FORCE PARTICIPATION RATES IN THE USSR, UNITED STATES, WEST GERMANY, AND JAPAN, 1980
(percent)

Age	*USSR*	*United States*	*West Germany*	*Japan*
15–19	36.95	46.30	54.65	19.00
20–24	83.45	78.10	77.20	73.10
25–29	96.20	80.55	76.95	73.65
30–34	97.65	79.60	77.80	72.70
35–39	98.10	80.05	78.65	77.15
40–44	96.90	79.90	77.60	80.00
45–49	93.15	77.20	74.00	80.10
50–54	82.95	72.50	67.95	77.60
55–59	66.20	64.60	55.15	70.05

SOURCE: *Economically Active Population: Estimates and Projections 1950–2025,* vol. 1 and 4 (Geneva: International Labor Office, 1986).

TABLE 5–12
PROJECTED NET CHANGES IN POPULATION OF WORKING AGES IN THE USSR AND SELECTED REPUBLICS AND REGIONS, 1980–2000
(thousands)

USSR, Republics, and Regions	*1980*	*1985*	*1990*	*1995*	*2000*
USSR (net increase in preceding 5 years)	11,788	3,649	2,341	2,932	7,240
Slavic republics					
RSFSR	4,956	–248	–1,081	–645	1,632
Belorussia	451	134	25	22	224
Ukraine	1,933	–39	–13	–262	184
Kazakhstan	1,148	772	670	674	891
Central Asia	2,291	2,106	2,130	2,569	3,278
Change in Kazakhstan and Central Asia as % of total working-age change	29.2	78.9	119.6	110.6	57.6
Change in Kazakhstan and Central Asia as % of change in Slavic republics	46.9	NA	NA	NA	204.4

NOTES: Males are ages 16–59; females, 16–54. Numbers are as of January 1. NA = not available.
SOURCE: Murray Feshbach, "The Soviet Union: Population Trends and Dilemmas," *Population Bulletin*, vol. 37, no. 3 (1982), p. 27.

constrain general productivity growth, and might well enhance it, undoubtedly it would make the task of augmenting deployable economic power less straightforward.

From the planners' standpoint, these potential additions to the labor force have the added disadvantage of residing in the "wrong" place. The great majority of the net increment in working age population will come in rural regions of Kazakhstan and the Central Asian republics. The areas of designated labor shortage, however, are resource-rich western Siberia and the Soviet Far East. There is no migration to

TABLE 5–13
URBAN POPULATION IN THE USSR AND ITS REPUBLICS, 1950–1980
(percent)

USSR and Republics	*1950*	*1960*	*1970*	*1980*
USSR	40.2	50.1	57.0	63.4
Slavic republics				
RSFSR	44.6	54.9	63.2	70.5
Ukraine	36.1	48.3	55.3	62.7
Belorussia	22.2	33.8	44.7	57.4
Moldavia	17.5	23.9	32.4	40.9
Baltic republics				
Estonia	49.6	57.8	65.6	70.5
Latvia	46.9	57.6	63.0	69.4
Lithuania	26.7	40.0	51.4	62.6
Transcaucasus				
Armenia	44.0	51.4	60.0	66.3
Azerbaydzhan	45.0	49.3	50.3	53.4
Georgia	36.3	43.4	48.1	52.4
Kazakhstan	39.3	44.5	51.2	54.9
Central Asia				
Kirgiziya	28.5	34.8	37.7	38.8
Tadzhikistan	27.3	34.4	37.3	34.3
Turkmenistan	38.3	47.4	47.8	47.8
Uzbekistan	30.7	35.2	36.5	40.2

NOTE: Data as of December 31 of each year.
SOURCE: Murray Feshbach, "The Soviet Union: Population Trends and Dilemmas," *Population Bulletin*, vol. 37, no. 3 (1982), p. 37.

speak of from the "southern tier" republics to those areas—nor has there been since the days of mass deportation. Indeed the rural populations of the Muslim republics are evidently reluctant even to move to cities in their own regions (where the European presence is stronger). In Tadzhikistan, for example, the fraction of the population in urban areas may have been lower in 1980 than it was in 1960 (table 5–13). Reorienting economic plans to make use of "low productivity" workers in "labor surplus" areas, from the standpoint of Soviet planners, must result in rates of growth for the socialist sector that are no more than "second-best."

Conclusions

The demographic dilemmas facing the Soviet Union are real enough. They may be expected to have an actual impact on Moscow's ability to augment and to deploy force. It would be exceedingly unwise, however, to read in these numbers a forecast of inexorable Soviet decline in the international arena. Soviet strategists appreciate that the correlation of forces is a complex quantity in which difficulties on one front may be offset by gains on another. The great military minds of the past have always respected the nonmaterial factors that may contribute to victory in struggle: strategy, morale, intelligence, and the like. We might do worse than to remember the words of Napoleon: "Even in war moral power is to physical as three parts out of four."

6

The CIA's Assessment of the Soviet Economy

Over the past four decades Western publications have recorded the progress of a vast and ongoing struggle: the struggle of its scholars and researchers to describe the Soviet economy. The earliest Western assessments of the USSR's economic performance and potential date practically to the foundation of the Soviet state. But before World War II such work was to a large degree episodic. It reflected the particular interests of individual specialists or the passing concerns of governments. Since World War II, by contrast, the struggle to describe the Soviet economy has been continuous, sustained, and comprehensive. The USSR's emergence from that war as a victor and great power, its early acquisition of nuclear weaponry, and its policy of hostility toward the United States and U.S. allies all lent urgency to an already challenging intellectual task: to understand and to explain the workings of an economic system fundamentally different from our own, and moreover masked from easy viewing by an officially enforced secrecy.

The postwar corpus of research and analysis on the Soviet economy includes work by every Western government and a huge volume of reports and studies from universities, institutes, and private organizations. For all the mass and diversity of these contributions, no specialist can fail to recognize the singular and central importance of the analyses produced by the U.S. government, principally under the

---

This chapter, previously unpublished, was written in 1990. It is reprinted as originally presented, as testimony before the Senate Foreign Relations Committee, July 16, 1990.

auspices of the Central Intelligence Agency.

The CIA's figures on Soviet economic trends are widely regarded as the most authoritative currently available. Classified CIA analyses inform U.S. policy makers, as well as counterparts in other governments with which they are shared. Unclassified CIA publications serve as basic reference sources on the Soviet economy in our universities, in our newspapers, and for the interested public. The CIA's longstanding effort to assess and to analyze the Soviet economy has involved many thousands of researchers, has tapped considerable talent, and has commanded enormous financial resources. It is probably safe to say that the U.S. government's attempt to describe the Soviet economy has been the largest single project in social science research ever undertaken.

In recent years, nevertheless, the CIA's assessment of the Soviet economy has come increasingly under question—and into criticism. The questions are fundamental; they concern the accuracy and reliability of the agency's most basic estimates of Soviet output and performance.

According to the CIA's assessment, the Soviet economy is slightly more than half as large as the U.S. economy; per capita consumption in the USSR is at roughly a third of the U.S. level; and aggregate output grew by about two-fifths between 1975 and the late 1980s.[1] To a growing chorus of critics, however, these estimates are all implausibly high. According to their assertions, the CIA has seriously and continuously overestimated the USSR's actual economic performance in a number of key respects. In the West, such criticisms are now expressed by a number of respected academic specialists on the Soviet economy, and by at least one senior official previously responsible for internal review of the CIA's intelligence assessments.[2] Since the advent of the glasnost campaign, such criticisms have also been voiced by some of the leading economists within the USSR itself.[3]

How is one to evaluate such criticism? Surely it would be unwise to minimize the practical and theoretical difficulties posed by any attempt to measure the product of a command socialist economy in a framework appropriate to a market-oriented order. It may not, in fact, be possible to prepare a single, unambiguous estimate in U.S. dollars for the value of goods and services produced each year in the USSR. Even under the best of circumstances the meaning of figures derived from such an exercise might not always be self-evident.

Insurmountable obstacles are one thing. Avoidable pitfalls are quite another. Some critics of the CIA's assessment of the Soviet economy argue that the agency's methods and practices are themselves sources of error, distorting the estimation process to such an extent that its results must necessarily be skewed. If such critics were correct, it would be possible, at least in principle, to improve the accuracy and reliability of estimates by correcting existing methodological biases or by relieving other influences directly distorting these computations.

A full review of methods and practices in the CIA's Soviet economic estimate is a task far beyond the scope of a brief note. This chapter is decidedly more modest: a comment upon a few aspects of the official, unclassified estimates of Soviet economic performance published regularly in such volumes as the CIA's annual *Handbook of Economic Statistics*. These statistics present the agency's summary reading of major trends in the Soviet economy. Insofar as these estimates are all generated internally by the CIA's Directorate of Intelligence, however, they directly and necessarily speak to its analytic approach, including its methods and practices.

These published estimates seem to raise as yet unanswered questions. Some of the series on the Soviet economy look decidedly anomalous; others appear to be internally inconsistent. A number of these problems happen to correspond with the broad proposition that the CIA's methods and practices result in an overestimate of Soviet output, consumption, and economic growth.

Use of Soviet Data

Three general sorts of problems can be identified in the CIA's estimates of Soviet economic performance. The first sort concerns the use of official Soviet statistics. Naturally, Soviet statistics are the basic data in the CIA's own estimates of Soviet output. But Soviet data must be handled with care if one hopes to use them for illumination. The shortcomings of these data are legion. All data involving official prices, for example, reflect preferences and priorities determined by the central government. This central fact frustrates outside attempts to measure value added by various economic activities in the USSR. No less significant is the subordination of the country's statistical system to its planning process. In a highly politicized setting, the performance of managers and administrators is judged by the correspondence between the quotas required of them by the plan and the figures they submit on their units' achievements. Under such circumstances, it is by no means obvious that errors in reporting should offset one another, as statistical authorities typically presume. Under these conditions, moreover, it is by no means clear that Soviet planners can make good use of one of statistical theory's most basic and useful techniques: the random sample survey.

U.S. government analysts are surely not incognizant of these qualifications. Yet frequently the CIA's analysis seems to take Soviet

TABLE 6–1
ESTIMATED MEAT AND MILK PRODUCTION IN THE UNITED STATES AND THE USSR, 1960–1988

	1960	*1970*	*1980*	*1988*
Meat				
Total output (million metric tons)				
United States	12.8	22.5	24.2	27.2
USSR	8.7	12.3	15.0	19.3
Per capita output (kilograms)				
United States	71	110	106	112
USSR	41	51	53	67
Percentage of world output				
United States	20	27	18	17
USSR	13	15	11	12
Milk				
Total output (million metric tons)				
United States	53.7	53.3	58.3	65.2
USSR	61.7	83.0	90.9	106.4
Per capita output (kilograms)				
United States	297	260	266	265
USSR	288	342	341	372
Percentage of world output				
United States	17	15	14	14
USSR	20	23	21	23

NOTE: Decimal places as printed in source.
SOURCE: Derived from U.S. Central Intelligence Agency, *Handbook of Economic Statistics, 1989* (Washington, D.C.: CIA, 1989), pp. 64, 44–45, 23.

figures at their face value, or to employ them after making only minor adjustments. Possibly U.S. analysts accept the general validity of Soviet data because that assumption facilitates the use of the complex econometric models they have devised. Analysts may hesitate to challenge Soviet data for fear of introducing arbitrary and inconsistent revisions into existing series. Whatever the justification, however, the practice can be expected to result in estimates of spurious precision and questionable accuracy.

The trouble with the approach is illustrated in table 6–1. This compares recent published CIA estimates of meat and milk output for the Soviet Union and the United States. Unlike estimates of national income or product, this comparison involves only volumes of physical output, and so circumvents the valuation problem. Moreover, agricultural commodities are relatively homogeneous by nature, and should at least in principle be more readily comparable than such things as consumer goods or capital equipment. One might therefore expect this comparison to be intrinsically more reliable than many others.

CIA estimates for Soviet meat production suggest that per capita output for the country as a whole was nearly a third higher at the end of the 1980s than at the start of the 1970s—a conclusion that would likely come as a surprise to many Soviet citizens who had lived through the period in question. By the late 1980s, according to these estimates, per capita production of meat in the USSR was roughly the same as that of the United States during the Eisenhower era. Such a reading is totally out of keeping with the general impressions of American and other Western visitors, to say nothing of Soviet researchers who have addressed this topic.[4]

The CIA figures on milk production are even more astonishing. By these numbers per capita output of milk in the late 1980s was about two-fifths greater in the Soviet Union than in the United States. Although press reports and visitor accounts for that period indicate that milk was no longer regularly available in food stores in various urban areas in the USSR, the CIA's estimates would have suggested that the country was fairly swimming in it. The USSR is represented to be far and away the world's largest producer of milk. By these indicators the USSR would seem to be not only a nuclear power but a dairy superpower as well.

Skeptical visitors who return from the USSR remarking upon the dissonance between CIA estimates and their own perceptions are sometimes cautioned that they have viewed only an unrepresentative portion of Soviet society or that the Soviet state allocates much of the country's resources to areas, activities, and projects that outsiders simply are not permitted to see. Such qualifications, however, can do little to explain dissonance for these particular estimates. If Western visitors are encouraged to visit showpiece locales, their impressions of the availability of meat, milk, and other luxury items should be unduly favorable. And while the Soviet state may routinely try to hide many of its most expensive efforts from public view, dissonance between estimate and impression in this case would seem to require vast secret stockpiles of milk and meat—perishable commodities not easily stored under current Soviet conditions.

The simplest explanation for this dissonance would be that the CIA was excessively credulous in accepting Soviet statistics and therefore failed to discount appropriately the upward bias in these official claims. Such credulousness may also help to explain why analysts apparently overlooked the existence of a serious and longstanding deficit in the Soviet state budget until after it was officially announced by Mikhail Gorbachev in 1988.

Budget data and other financial figures, it is now officially admitted, were deliberately doctored and falsified for decades. Yet such official efforts at deception were by no means impenetrable. Far from it: almost a decade ago a study by an émigré researcher identified this secret budget deficit, and calculated its magnitude.[5] That study managed to uncover a hidden feature of the Soviet economy through the painstaking review of Soviet sources and an attention to statistical discrepancies. Unfortunately, the same objective skepticism does not seem to be characteristic of the CIA's treatment of some Soviet data.

Valuation of Soviet Output

A second set of problems concerns the method by which the CIA values Soviet output. The CIA's methodology for computing aggregate Soviet GNP and its component parts has been described in general terms in the open literature, although the details of its procedures remain classified secrets. Even without delving into the fine points of the procedure, however, one may observe that the results it generates look at times to be anomalous, if not actually internally inconsistent.

Until 1989, for example, published CIA estimates indicated that productivity in the Soviet economy had been falling steadily throughout the 1970s and through most of the 1980s.[6] Yet at the same time agency estimates indicated a steady, even respectably rapid increase in per capita consumption in the USSR (table 6–2). Over the 1960s and 1970s, CIA estimates suggested that while aggregate factor productivity had on balance declined, the levels of per capita consumption had come close to doubling.

It is not immediately obvious how one would reconcile such disparate readings of Soviet performance. If long-term productivity trends in an economy were negative, one might nevertheless in theory increase per capita consumption substantially through radical reallocation of resources toward the consumer. According to CIA estimates, however, the share of consumption in the Soviet economy was no higher in 1980 than it had been in 1960. (By their numbers, in fact, it was actually said to be slightly lower.) One might try to reconcile the paradoxes reported by supposing a concerted structural shift for nonconsumption resources into investment and away from all other purposes. Yet for such an arithmetic solution to be plausible, one would have to posit a vast redirection of resources away from the Soviet armed forces between 1960 and 1980—a proposition completely at odds with what has been reported in the West (and now grudgingly conceded in Moscow) about the USSR's military buildup during that period.

TABLE 6–2

ESTIMATED CHANGES IN AGGREGATE FACTOR PRODUCTIVITY, PER CAPITA CONSUMPTION, AND SHARE OF CONSUMPTION IN THE SOVIET ECONOMY, 1961–1987

	1961–65	*1966–70*	*1971–75*	*1976–80*	*1981–85*	*1987*
Aggregate factor productivity (percentage per year)	0.3	0.8	–1.2	–1.3	–1.1	–1.9
Per capita consumption	2.5	5.0	3.0	2.0	NA	0.7
Consumption as percentage of GNP	54 (1960)	51 (1970)	53 (1975)	53 (1980)	54 (1985)	54

Total change 1961–80	
Total factor productivity	–7%
Per capita consumption	+85%
Share of consumption in GNP	–1 percentage point

SOURCE: U.S. Central Intelligence Agency, *Handbook of Economic Statistics, 1988* (Washington, D.C.: CIA, 1988), pp. 62, 63, 65.

A simpler explanation for these ostensibly conflicting estimates would be methodological error. If the CIA's method had relied upon systematically inadequate deflators for its Soviet series—if, in other words, inflation in the USSR were not fully taken into account—resultant computations could simultaneously exaggerate the decline in Soviet productivity and the increase in the country's levels of per capita consumption.

In Soviet-style systems, where prices do not necessarily reflect scarcity and where markets cannot be presumed to be moving toward equilibrium, inflation is rather difficult to measure. "New" goods replacing older models sometimes differ only in price. The quality of given products may actually decline over time. Moreover both rationing and shortage regularly serve as adjustment mechanisms for coping with excess demand. If these phenomena are not adequately discounted, deflators will necessarily overestimate an economy's output and inputs alike.

Over decades, one should remember, even a relatively small annual underestimate of a phenomenon can introduce consequential distortions into economic estimates. Such distortions could become all the more pronounced if there were an acceleration of the phenomenon that was being inadequately measured. It is commonly thought that the pace of inflation in the USSR has quickened during the past fifteen years, since the mid-1970s.

Robustness of Estimates

A third problem evident in the estimates pertains to what might be called their robustness. In the social sciences, as in the natural sciences, hypotheses and calculations are ordinarily exposed to tests in which they can be challenged and thereby potentially refuted. Theories and results can be validated only through the repeated attempt to have them falsified. The CIA's estimates for Soviet economic performance do not appear to have been rigorously exposed to such external challenges. While it is admittedly often much more difficult in economics than in the physical sciences to frame an experiment that can conclusively falsify a result, a variety of data, estimates, and events familiar to specialists in Soviet studies and allied disciplines would seem implicitly to challenge the reliability and accuracy of the CIA's recent series on Soviet economic performance. How recent CIA estimates are to be reconciled with some of these implicit challenges is not immediately obvious.

Mortality data provide an example of such implicit challenges. As is now well known, age-specific mortality for many Soviet cohorts rose substantially between the mid-1960s and the mid-1980s. So pronounced were these increases in death rates that life expectancy at birth registered a decline for both men and women.[7] (Such demographic trends, one may note, are inherently easier to measure and validate than most economic trends.) By the late 1980s, according to data compiled by the World Health Organization, the USSR's mortality levels were similar to those of some Latin American countries. By one measure—age-standardized death rates (WHO's European Model)—mortality for both men and women around 1987 was apparently higher in the USSR than in such places as Argentina, Chile, and even Mexico (table 6–3).

A population's mortality level bears some relation to its economic circumstances. Other things being equal, one would expect a population in which per capita consumption was increasing to enjoy general improvements in life expectancy as well. To the extent that human capital figures in the economic process, deterioration of public health—and especially deterioration in the health of the working-age population—should constrain output.

None of this is to say that per capita consumption or per capita output today must necessarily be lower in the USSR than in those Latin American countries with better levels of public health. It would

TABLE 6–3
MORTALITY RATES IN THE USSR AND SELECTED LATIN AMERICAN COUNTRIES, 1985–1988
(deaths per 100,000 population)

Country	*Male*	*Female*
USSR (1987)	1,525	874
Argentina (1985)	1,241	810
Chile (1987)	1,189	797
Costa Rica (1988)	1,183	768
Mexico (1986)	1,200	872
Uruguay	1,229	783

NOTE: Figures are rounded to nearest digit. The age structure is from the European Model of the World Health Organization.
SOURCE: World Health Organization, *World Health Statistics Annual 1989* (Geneva: WHO, 1989), pp. 388–94.

seem incumbent upon those assessing Soviet economic performance, however, to recognize and to deal with the dramatic disparity between these mortality data and the agency's own economic estimates for the countries in question. According to the 1989 edition of the *Handbook of Economic Statistics,* for example, Soviet per capita output in 1988 was more than three and a half times that of Argentina and more than four times that of Mexico.[8] No explanation of this discrepancy has been volunteered.

Another implicit challenge to the estimate centers upon the reproducibility of its results. Nonclassified research adhering ostensibly to the same general approach in assessing output of pre-1989 Soviet-bloc economies has resulted in estimates widely divergent from the CIA's own. The 1989 *Handbook of Economic Statistics* indicates its estimates for the USSR and Eastern Europe are adjusted for purchasing power to improve comparability for Western economies.[9] To date, however, the most extensive and detailed attempt to assess international purchasing power parities is the ongoing International Comparison Project (ICP). The ICP, currently coordinated and cofinanced by the United Nations Statistical Office, has been under way for more than twenty years and now involves more than sixty countries, all of whose governments have actively cooperated with the ICP's investigations. Several Eastern European countries, then Communist, participated in successive phases of the ICP's study. In its phase V study (for the year 1985), the ICP attempted to standardize the levels of per capita output in these countries against those of other countries in Europe.

CIA and ICP estimates of relative per capita output for various

TABLE 6–4
ESTIMATED PER CAPITA OUTPUT FOR SELECTED EUROPEAN COUNTRIES AND THE USSR, 1985
(West Germany = 100)

Country	*UN Estimate (GDP)*	*CIA Estimate (GNP)*	*CIA/UN*
West Germany	100	100	1.00
France	94	97	1.03
United Kingdom	90	91	1.01
Italy	89	91	1.02
Hungary	42	63	1.48
Yugoslavia	40	49	1.23
Poland	33	55	1.64
USSR	NA	64	NA

NOTES: UN and CIA estimates both attempt to adjust for purchasing power parities.
NA = not available.
SOURCES: Derived from United Nations Statistical Commission and Economic Commission for Europe, *International Comparison of Gross Domestic Product in Europe 1985* (New York: UN, 1988), p. 5; and CIA, *Handbook of Economic Statistics, 1989*, pp. 30–31, 44–45.

European countries in 1985 are interesting to compare (table 6–4). For countries in Western Europe, CIA and ICP rankings are close. The slight differences between their results might possibly be explained by the slightly different definitions of "national output" in the two series. (The CIA looks at gross national product, while the ICP estimates gross domestic product.) For Eastern Europe, however, the CIA's estimates are consistently and often dramatically higher than those of the ICP. In its relative ranking the CIA's per capita output estimates are almost 50 percent higher than the ICP's for Hungary and nearly two-thirds higher for Poland. These discrepancies cannot be explained away by national accounting procedures. Gross national product can be sharply higher than gross domestic product only if a country is receiving substantial remittances from workers employed or capital being invested overseas; such conditions did not obtain in these Communist countries in the mid-1980s.

The ICP's phase V adjustments for purchasing power included an attempt to reflect differences in the quality of output among countries.[10] Exactly what the CIA did is unclear since its method and procedure are not publicly explained in detail.[11] CIA researchers, however, have not commented upon the discrepancy between their

Soviet-bloc purchasing power estimates and those of this major international study.

The CIA's estimates for Soviet-bloc economies would seem to be challenged further by revelations and research since the revolutions of 1989 in Eastern and Central Europe, and particularly by events in East Germany. In the *Handbook of Economic Statistics 1989,* per capita output in 1988 in East Germany is placed at roughly seven-eighths of the West German level. Since the collapse of East Germany's Communist regime and the opening of that formerly closed society, it has become apparent that such an estimate would be regarded as utterly fanciful by many ordinary East Germans. With virtually no exceptions, press reports since the opening of the Berlin Wall have indicated that East Germans generally regard their living standards, and output levels, to be far lower than those in West Germany.

Even before "die Wende," however, CIA estimates of East German economic performance were seriously at variance with those produced in West Germany. For a variety of reasons, West German researchers have paid close attention to the East German economy over the past four decades. The German Economic Research Institute (DIW) has been studying the East German economy intensively since the 1950s; its work is considered quite authoritative within Germany, insofar as the Bundestag commissions it to prepare estimates and assessments on the economic situation in the German Democratic Republic. According to a study presented in 1987, output per worker in the early 1980s was only half as high in East Germany as in West Germany; even allowing for its higher labor force participation rates, per capita output in the GDR was said to be no more than three-fifths the FRG's level.[12] Since those estimates were completed, much more information has become available. West German researchers have been revising their estimates—downward. The DIW's 1990 figures place GNP per capita in 1989 at only 48 percent of the West German level[13]—barely half what CIA figures would suggest.

DIW's estimates may be revised still further in coming months and years. They should not be taken as the final word on economic performance in the German Democratic Republic during the regime's final years. But the enormous gap between its current estimates and the CIA's estimates has implications for the entire set of estimates on Soviet-bloc economies that the agency regularly prepares. To date, the CIA has not publicly addressed these implications.

Benefits of Review

In sum, even without examining classified materials, numerous questions can be raised about the accuracy and reliability of the official

U.S. assessment of the Soviet economy. These questions are sufficiently central that they might properly invite more intensive and comprehensive review. Only through full, thorough, and impartial discussion of methods and data can these estimates be definitively critiqued—or satisfactorily substantiated. In the absence of such efforts, doubts about the validity of the current U.S. estimate of Soviet economic performance cannot be effectively dispelled.

Whatever such a review might determine, there is, at a minimum, reason to wonder whether the U.S. government has not significantly mismeasured Soviet economic performance. It might therefore be prudent to consider some to the potential implications of mismeasurement. The CIA's assessment of the Soviet economy, after all, is intended to be policy research. U.S. policy has been shaped by these assessments—and indeed is supposed to be.

A systematic and continuing overestimate of Soviet economic output and growth would bear upon at least three areas of policy concern. The first relates to the USSR's military posture. Estimating the actual level of military expenditures in the USSR is no easy matter. Western governments, however, believe they have a fairly good idea of the Soviet military apparatus's offensive and defensive capabilities. If their assessments for Soviet military manpower, hardware, materiel, and logistical support facilities are roughly accurate, a smaller Soviet economy would necessarily mean a greater military burden than has to date officially been suggested in Washington.

CIA estimates currently indicate the Soviet economy to be highly militarized. With a military burden reckoned at 15–17 percent of national output, the USSR would seem to be a country operating on a semiwar footing (like the European powers in the years just before World War II). Alternative denominators for the Soviet economy, by contrast, might place the USSR's military burden at 20 to 30 percent of national output during the 1970s and 1980s. Such a level would be one consonant with full war mobilization; in 1942, for example, military expenditures equaled roughly 21 percent of national output in the United States.[14]

The distinction is not trivial. Understanding an adversary's intentions is no less essential than understanding his capabilities. The USSR's military burden is, among other things, a measure of the intentions and outlook of the Soviet leadership. If Soviet authorities had committed their economy to financing something like a full war mobilization that lasted for decades, that decision in itself would be extraordinarily significant. If American assessments have underestimated the Soviet military burden, they would have correspondingly misassessed Soviet government intentions for decades.

TABLE 6–5
ESTIMATED REAL ECONOMIC GROWTH IN THE USSR AND SELECTED COUNTRIES, WESTERN EUROPE, 1981–1988
(percent per year)

Country	*1981–85*	*1986–88*	*1981–88*
USSR	1.9	2.3	2.0
European Community	1.5	2.9	2.0
France	1.5	2.5	1.9
Italy	1.6	3.2	2.2
United Kingdom	1.8	3.7	2.5
West Germany	1.3	2.5	1.7

NOTE: Estimates are for GNP for the USSR and for GDP for Western European countries.
SOURCE: U.S. Central Intelligence Agency, *Handbook of Economic Statistics, 1989* (Washington, D.C.: CIA, 1989), p. 33.

A second area relates to Gorbachev's campaign for perestroika. If Soviet output and growth have been lower than officially supposed, the motivations for this restructuring might be different from those commonly presumed.

Recent CIA estimates suggest that the Soviet economy grew roughly 2 percent a year during the 1980s. Such a pace of growth would widely be regarded as unexceptional. By these estimates, in fact, the Soviet economy would have been growing just as rapidly during the 1980s as the economies of the European Community and more rapidly than the French or West German economies[15] (table 6–5).

If the Soviet economy was indeed muddling along with slow but steady growth during the "years of stagnation" before Gorbachev's rise to power, the notion that perestroika and attendant international policies were initially propelled by "new thinking" might seem quite plausible. If, conversely, Soviet circumstances were more dire than CIA assessments indicated, one might not need to posit new thinking to account for far-reaching policy changes in the USSR after the Gorbachev accession.

A final area of implications relates to the future. While the CIA's assessment depicts a Soviet economy that is manifestly inefficient and burdened by military expenditure, its basic figures nevertheless would seem to point to sustainability and continuity. Steady if lackluster growth is not ordinarily understood to presage systemic crisis. In these estimates, slowdown, not breakdown, would appear to be the principal problem looming on the economic horizon. On the basis of existing CIA figures alone, Western policy makers would have to judge the probability of a major dislocation in the Soviet economy in the foreseeable future to be rather low.

Time alone will tell the fate of the Soviet economy. Alternative readings of Soviet economic performance, however, do exist. Some of these could be easily reconciled with contemporary Soviet accounts, and foreign reports, of pervasive shortage, systemic decay, and uncertainty in the USSR's economy. Such assessments might leave policy makers less surprised, and less unprepared, if they do eventually find themselves forced to cope with events precipitated by economic instability in the USSR—or by the collapse of the Soviet economy itself.

7

Poverty in South Africa

Few readers would be surprised to hear that great pockets, if not entire strata, of South African society suffer from material want. Many more, however, might be surprised by the proposition that the contemporary South African political system is not only extremely sensitive to the problem of poverty but committed to eradicating it. Yet this is the case. It may be said, with only slight exaggeration, that South Africa's postwar evolution was propelled by the quest to solve the poverty problem, and that the modern South African state represents the machinery that was meant to bring the solution into being.

To no inconsiderable degree, the postwar apartheid policy of the Republic of South Africa (RSA) is a political response to the problem of poverty. To be sure, it is a highly specialized response to a particularly delimited vision of the problem. For the first half of the twentieth century, to talk in South Africa of poverty was to talk of *white* poverty, to discuss the plight of what were called the "poor whites." Poor whites, it was generally agreed, accounted for a substantial fraction of South Africa's population of European descent. Their poverty was real enough. As documented in the five-volume Carnegie Commission report, *The Poor White Problem in South Africa* (1932), and elsewhere, poor health, malnutrition, illiteracy, destitution, and vagrancy figured in the lives of a sizable minority of the country's whites. White poverty did not arrive with the Great Depression, or disappear at its termination. On the eve of World War II, Karl Bremer, then a member of the South African parliament, reported that 15 percent of the rural population did not "get enough food . . . to ward off the prevalent diseases,"[1] and a survey by the Union Department of Public Health concluded that more than a quarter of South Africa's white schoolboys had no milk at all in their diets.[2]

The poor whites in question were overwhelmingly Afrikaans-speakers. It was no political secret. In 1943, for example, an electioneering pamphlet for the National party declared that "the Afrikaner nation is the poorest element in the white population, and is even poorer than the Indians."[3] It was this distress that the Afrikaner nationalists purposefully set out to relieve once they came to political power in 1948. They did so not only by devising new policies but by making use of existing instruments of colonial policy. Their program included "racially" specific wage preferences in favor of whites, severe restrictions against "native" laborers in competing for work and business, preferential quotas and set-asides for whites, and a systematic expansion of government jobs. In a sense, apartheid may be seen as a vast and ambitious antipoverty program: as the earliest, and to date the most audacious, of the affirmative action programs to be adopted in the postwar era by Western governments in the name of uplifting a group deemed to be disadvantaged.

As an antipoverty program for Afrikaners, apartheid looks to have been quite successful. White poverty by 1948 was on the decline, but it is now for all intents and purposes a thing of the past. Moreover, the social gap between English- and Afrikaans-speaking whites seems to have narrowed considerably. According to S. J. Terreblanche of Stellenbosch University, per capita income for Afrikaans-speaking whites was less than half that for English speakers in 1946.[4] In 1980, by his estimate, it was up to about 70 percent—and one may suspect that some portion of the remaining differential can be explained in terms of differences in fertility, age structure, and living costs between the two populations.

For a diversity of reasons, the question of poverty in South Africa has evoked renewed curiosity and concern in the 1980s. A comprehensive assessment of poverty in South Africa, however, requires evaluation of the plight and progress of *all* population groups in the country—White, Asian, Coloured, and Black—and not merely the ones who are meant to benefit especially from government action. (Although some readers will object to the use of the South African government's quadripartite race terminology here and elsewhere in this chapter, the taxonomy is pertinent precisely because apartheid laws invested these mandatory assignations with so much consequence.)

Rhetorical allusions to poverty in South Africa are staple features of the international campaign to stigmatize and to isolate the country. But new attention to the problem has also come from within South

Africa itself, where it has prompted a considerable volume of scholarly and academic work. In 1984 a Second Carnegie Inquiry into Poverty and Development in Southern Africa convened under the direction of Francis Wilson of the University of Cape Town; its findings and recommendations are informed by the more than three hundred papers it has commissioned and assembled.

It is not nearly as easy to map out the contours of poverty in contemporary South Africa as one might think. For one thing, South Africa's system of social and economic statistics is strangely primitive for a country of its technological, managerial, and administrative attainment. Entire categories of data pertaining to the material well-being of the overwhelming majority of South Africa's population—the blacks—are either not readily available or not collected at all. (Such omissions and lacunae may in themselves be revealing about the nature of poverty in South Africa.) For another thing, poverty does not really lend itself to unambiguous classification. Everyone may know poverty upon seeing it, but the condition is not so easy to define. Poverty is not a single, homogeneous quantity. Its manifestations may be as different and diverse as the individuals in whom it is represented. For this reason alone, the assessment of poverty tends to be somewhat arbitrary and incomplete, even under the best of circumstances. Nevertheless, we may go far toward an understanding of material poverty in South Africa today by reviewing available information in four broad areas: population, health, nutrition, and literacy.

Population

The distribution of population—by racial designation and geographic location—gives an initial indication of the contours of poverty in South Africa. According to the 1985 census, the total population of the Republic of South Africa on March 5 of that year was about 23.4 million. Whites were said to account for slightly less than one-fifth of this total, Coloureds just under one-eighth, and Asians a little less than one-thirtieth; the majority of South Africans, five-eighths of the population counted, were Blacks. These figures, however, are misleading. By the estimate of the quasi-official Human Sciences Research Council (HSRC), the 1985 census undercounted South Africa's population by more than four million people, and nearly 90 percent of those missed are thought to have been Blacks.[5] Moreover, the 1985 census explicitly excluded the populations of those homelands (Bantustans) that had been placed in formal independence (Transkei, Bophuthatswana, Venda, and Ciskei—the so-called TBVC countries), although these people are in many senses a definite part of South

TABLE 7–1
POPULATION BY RACIAL DESIGNATION IN SOUTH AFRICA, 1985

Racial Designation	*Population (thousands)*	*Total Population (%)*
Official RSA Census Count, March 5, 1985		
White	4,569	19.6
Asian	822	3.5
Coloured	2,833	12.1
Black	15,163	64.8
Adjustment by Human Sciences Research Council for Estimated Net Underenumeration		
White	4,947	17.8
Asian	861	3.1
Coloured	2,862	10.3
Black	19,052	68.7
Adjustment by HSRC plus Development Bank of Southern Africa Estimates for Independent Homelands		
White	4,947	14.7
Asian	861	2.6
Coloured	2,862	8.5
Black	25,006	74.3
Total	33,676	100.1

NOTE: Percentage totals may not add precisely because of rounding. The exact date of Development Bank of Southern Africa estimates for independent homelands population is not given. Independent homelands estimates exclude a small number of persons designated as Coloured, Asian, or White residing within those boundaries. Population totals and percentages may be slightly affected by these omissions and oversights.
SOURCES: Derived from Republic of South Africa, *Population Census 1985: Geographical Distribution of the Population with a Review for 1960–1985* (Pretoria: Central Statistical Services, Report 02-85-01, 1986), p. xix, and Carole Cooper et al., *Race Relations Survey 1985* (Johannesburg: South Africa Institute of Race Relations, 1986), p. 1.

Africa. By the perhaps overly precise estimate of the Development Bank of Southern Africa (Pretoria no longer collects statistics on the independent homelands), the TBVC population in 1985 was 6.03 million.

Adjusting for the census undercount and the areas left uncounted significantly alters the demographic picture of South Africa (table 7–1). By this revised reckoning, Whites were not a fifth but rather a seventh of the South African populace in 1985; Blacks made up nearly three-

TABLE 7–2

ESTIMATED GEOGRAPHICAL DISTRIBUTION OF BLACK POPULATION IN SOUTH AFRICA, 1960–1985

(thousands)

	White Areas	*Self-governing National States*	*Independent Homelands*	*Total*
1985				
Urban	6,476	1,303	691	8,470
Rural	4,481	6,792	5,372	16,645
Total	10,957	8,095	6,063	25,115
Urban (%)	59.1	16.1	11.4	33.7
	1960	*1970*	*1980*	*1985*
1960–1985				
Black population in urban White areas (%)	29.2	28.3	27.3	25.9
Black population in homelands (including independent homelands)	39.4	47.2	53.8	56.7
Black population, urban (%)	31.8	33.1	33.4	33.7

NOTE: Population totals for 1985 for White areas and self-governing national states reflect HSRC adjustments for estimated net undercount; urban and rural estimates derived according to the assumption that the ratio of net underenumeration for urban and rural areas was constant within White areas and within self-governing national states, respectively. The 1985 figures for independent homelands are derived from revised estimates of the Development Bank of Southern Africa for total population and proportion of total population in urban areas. The 1980 estimates are adjusted by HSRC estimates of undercount in White areas and self-governing national states, and apply to independent homelands their proportion of population urban; this procedure may result in a slight overstatement of the urbanization level of the Black population in South Africa for 1980.

SOURCES: For independent homelands: "Supplement-Black Development in the Independent National States: Tables" *Development Studies Southern Africa*, October 1982, p. 97; Development Bank of Southern Africa, unpublished data. For RSA: Republic of South Africa, *South African Statistics 1986* (Pretoria: Central Statistical Service, 1986); idem., *Population Census 1985: Geographical Distribution of the Population with a Review for 1960–1985* (Pretoria: Central Statistical Service, 1986, 02-85-01); Charles Simkins, "The Demographic Demand for Labour and Institutional Context of African Unemployment in South Africa: 1960–1980," SALDRU Working Paper 3 (Cape Town, August 1981).

TABLE 7–3
ESTIMATED PER CAPITA REAL DISPOSABLE INCOME, BY RACIAL DESIGNATION AND LOCATION IN SOUTH AFRICA, 1980 AND 1983
(Rand at 1983 prices)

	Estimated per Capita Disposable Income		*Ratio (white = 100)*	
Group	1980	1983	1980	1983
White	6,637	6,242	100	100
Asian	2,112	2,289	32	37
Coloured	1,525	1,630	23	26
Black				
Metropolitan areas	1,337	1,366	20	22
Nongrowth areas	383	388	6	6
Total Black	723	740	11	12

NOTE: "Metropolitan areas" is a Bureau of Market Research classification, based upon areas delineated in the South African census. "Nongrowth areas" is also a BMR classification, pertaining principally to the rural areas of the self-governing national states and the independent homelands in South Africa.
SOURCE: H. de J. van Wyk, *The Economic Importance and Growth of Selected Districts in the RSA and TBVC Countries in 1985* (Pretoria: University of South Africa, Bureau of Market Research, Report 114, March 1984).

quarters of the entire population and accounted for about seven-eighths of the non-Whites.

Table 7–2 outlines the distribution of the Black population in South Africa. In 1985 (after adjusting for the undercount) a little more than two-fifths of Black South Africans lived in territories mandated to Whites under the Group Areas Act (White areas); about a third lived on Bantustans within the boundaries of the RSA; and a little less than a quarter lived in the Bantustans that had been formally separated from the RSA (independent homelands).

Dramatic differences in urbanization characterize these three regions. About three-fifths of the Blacks in White areas live in cities. By contrast, less than one-seventh of the Black population on the Bantustans live in areas classified as urban. A generation and more of "separate development" notwithstanding, three-quarters of South Africa's urban Blacks today live in White areas.

The proportion of the Black population residing in urban White areas, however, has not been on the increase. Over the past quarter-century a profound and consequential shift in the distribution of South Africa's Black population has taken place. The urban sector in White areas has been, and still is, the principal locus of economic opportunity in South Africa. In the early 1980s, by the estimate of the University of South Africa's Bureau of Market Research, per capita disposable income was about three and a half times as great for Blacks in metropolitan areas as for those in nongrowth areas, principally Bantustans (table 7–3). Yet between 1960 and 1985 the share of South Africa's Black population living in urban white areas actually *fell.* Over the same period the proportion of Blacks living on Bantustans increased tremendously: a jump from about 39 percent to almost 57 percent. This massive trek away from economic opportunity was neither spontaneous nor natural. It was instead something like a forced march, supervised under the recently dismantled apparatus of "influx control" through such instruments as passbooks and involuntary relocations.

As the result of this most unusual pattern of migration, the urbanization of the Black population came to a virtual halt. The Black population has grown rapidly in recent decades, and natural increase in the cities has paralleled the general trend. But the proportion of Blacks in urban areas has gone almost unchanged in recent decades. In 1970, by one calculation, 33.1 percent of the Black population of South Africa resided in cities; in 1985, including TBVC, the corresponding figure was 33.7 percent. Urbanization in effect has been made to cease—and at a level whereby nearly two-thirds of the Black population remained rural. The urbanization level for Blacks is dramatically lower than for South Africa's other non-White groups. In 1985 more than three-quarters of the Coloured population of South Africa lived in cities, and less than 7 percent of the Asian population was in the countryside.[6]

Of all the population groups in South Africa, the rate of natural increase for the Blacks has been the most rapid in recent decades. Given the low level and stagnant pace of urbanization for blacks, their general process of urbanization in South Africa has effectively been arrested for a quarter of a century. According to South African statistical yearbooks, the proportion of South Africans in urban areas rose between 1960 and 1985—from about 47 percent to about 56 percent. Such computations rely upon the exclusion of the TBVC population from South Africa for 1985. If these people are included in the tally, using Development Bank of Southern Africa numbers, South Africa's urbanization ratio may have actually *declined* slightly between 1980 and 1985. That is to say, South Africa appears to have grown increasingly *rural* in recent years.

Along with the People's Republic of China, South Africa is one of the few places in the modern world where restrictive measures, vigilantly enforced, have successfully prevented urban migration for rural majorities. Such barriers have the predictable effect of preserving or even widening differences in material well-being between urban and rural populations. Material living standards are typically expected to be higher in urban than in rural areas in the modern world, yet the disparities associated with policies that confute voluntary domestic movement are particularly striking. In the early 1980s, for example, a World Bank team estimated the difference in life expectancy between city and country in China to be about twelve years[7]—far greater than the estimated gap for India, Indonesia, or any other poor and populous country. In South Africa, government policy has apparently also imputed special consequence to the fact of rural life for the vast majority of the population.

Health

If health is an imperfect proxy for material well-being, it is a good one nonetheless. Within and among societies, health and poverty correspond, often closely. In part, this is because good health comes close to being a universally desired attribute and because so many of the things that can make good health possible—proper nutrition, good hygiene and housing, medical care, education, and the like—are associated with the standards of living that affluence can bring.

The most basic measure of health is the death rate, and the most succinct and unambiguous measure of mortality for an entire population is its life expectancy at birth. Table 7–4 presents official estimates of life expectancy at birth for the three population groups in South Africa for which such numbers may be reliably computed: Whites, Asians, and Coloureds (table 7–4).

Unsurprisingly, life expectancy for both men and women is today highest for Whites. Around 1980, life expectancy for Whites was about five years higher than for Asians and about twelve years greater than for Coloureds. These differences in life chances speak directly to other differences in material circumstances.

What may be somewhat surprising is the record of health progress for the different population groups. Contrary to what might be expected, health gains for the White population between 1951 and 1980 were minimal. Between 1951 and 1970, life expectancy at birth for the White population as a whole increased less than a year and virtually halted for men; during this period life expectancy for South Africa's White adults actually *declined*. Paradoxically health progress among Whites seemed to slow precisely at the time that the apartheid policy was

TABLE 7–4

LIFE EXPECTANCIES AT BIRTH BY RACIAL DESIGNATION IN SOUTH AFRICA, 1935–1981
(years)

Racial Designation	*1935–37*	*1945–47*	*1950–52*	*1959–61*	*1969–71*	*1979–81*
Males						
White	59.0	63.8	64.6	64.7	64.7	66.6
Asian	NA	50.7	55.8	57.7	59.2	62.3
Coloured	40.2	41.7	44.8	49.6	48.9	54.3
Females						
White	63.1	68.3	70.8	71.7	72.4	74.2
Asian	NA	49.8	54.8	59.6	63.2	68.4
Coloured	40.9	44.0	47.8	54.3	55.8	62.6
Difference in Years from White Life Expectancy						
Males						
Asian	NA	13.1	8.8	7.0	5.6	4.3
Coloured	18.8	22.1	19.8	15.1	15.9	12.3
Females						
Asian	NA	18.6	15.3	12.0	7.2	5.9
Coloured	22.2	24.3	22.3	17.4	16.5	11.7

	1935–1952	*1950–1971*	*1969–1981*
Changes in Life Expectancy in Years			
Males			
White	5.6	0.1	1.9
Asian	NA	3.4	3.1
Coloured	4.6	4.1	4.7
Females			
White	7.2	1.6	1.8
Asian	NA	8.4	5.2
Coloured	6.9	8.0	6.8

NA = not available.
SOURCE: Republic of South Africa, *South African Life Tables 1979–1981* (Pretoria: Central Statistical Services, Report 02-06-03, 1985).

striving to confer special benefits upon the White population and to uplift the most disadvantaged elements within the White group.

Whatever the reasons for this comparatively poor performance,

TABLE 7–5
INFANT MORTALITY RATES BY RACIAL DESIGNATION IN SOUTH AFRICA, 1940–1985
(deaths per 1,000 live births)

Racial Designation	*1940*	*1950*	*1960*	*1970*	*1975*	*1980*	*1985*
White	50.1	35.7	29.6	21.6	20.1	13.1	9.7
Asian	89.9	68.5	59.6	36.4	34.7	24.4	15.6
Coloured	156.9	134.3	128.6	132.6	104.0	60.7	40.4
Infant Mortality Ratios (white = 100)							
White	100	100	100	100	100	100	100
Asian	179	192	201	169	173	186	161
Coloured	313	376	434	614	517	463	416

Racial Designation	*1940–60*	*1940–70*	*1970–85*	*1975–85*
Annual Rate of Decline (%)				
White	–2.6	–2.8	–5.2	–7.0
Asian	–2.0	–3.0	–5.5	–7.7
Coloured	–1.0	–0.6	–7.6	–9.0

NOTE: Data on black infant mortality are not comprehensively collected by South African statistical services.
SOURCES: Republic of South Africa, *South African Statistics 1986* (Pretoria: Central Statistical Service, 1986), pp. 3.5–3.7, and *Bulletin of Statistics*, December 1986, p. 1.4.

life expectancy for South African Whites was distinctly lower than for most Western populations by 1980. On the basis of life expectancy figures alone, one might have guessed South Africa's Whites in 1980 to have been a population from Eastern Europe, or perhaps the Caribbean. Life spans in 1980 were about the same for South African Whites as for Jamaicans.[8]

Asians and Coloureds, conversely, enjoyed substantial health progress during the first three decades of apartheid. Between 1951 and 1980, life expectancy at birth is officially estimated to have risen about a decade for Asians and about a dozen years for Coloureds. Dramatic though the remaining differences may be, between 1951 and 1980 the life expectancy gap narrowed by more than five years between Whites and Asians and by more than nine years between Coloureds and Whites.

The infant mortality rate (the number of babies to die in their first year of life per thousand born) is another measure of health in a society. The rate is considered significant for a number of reasons. Infant mortality has a strong influence on general levels of life expectancy. It reflects the well-being of an especially vulnerable group. And it is a phenomenon that can be systematically attacked through concerted government effort, even when income levels are low.

Table 7–5 shows the official figures on infant mortality in South Africa for Whites, Asians, and Coloureds. In 1985 infant mortality was 60 percent higher for Asians than Whites and more than four times as high for Coloureds as for Whites. By these numbers the 1985 Asian infant mortality rate was about the same as the White rate of the late 1940s.

The 1985 infant mortality rates for these population groups can be compared with contemporary rates in the rest of the world, using U.S. Census Bureau figures and estimates.[9] For the White population, infant mortality was about the same as in the United States and Belgium. The Asian infant mortality rate was similar to those given for Barbados and Puerto Rico. The Coloured rate was at roughly the same level imputed to such places as Mexico and the West Bank in the Holy Land.

Between the mid-1960s and the mid-1970s there was little improvement in infant mortality for Whites. The White infant mortality rate, once comparatively low, lagged against those of developed Western nations by the late 1970s. By such comparisons deaths from diseases associated with poverty were particularly high. In 1977, for example, the pneumonia death rate for White South African infants was more than twice as high as in England and Wales, four times as high as in the United States, and seven times as high as in Sweden.[10]

The pace of the infant mortality decline stepped up sharply for Whites in the late 1970s and the early 1980s. It was even more rapid for Asians and Coloureds. The recent pace of improvement for Coloured infants is particularly noteworthy. Between 1950 and 1970 virtually no decline was registered in infant mortality rates for the Coloured population. Between 1970 and 1985 the Coloured infant mortality rate fell more than two-thirds. Between 1975 and 1985 the Coloured infant mortality rate dropped 9 percent a year. Hardly any other population on record has to date enjoyed such a rapid and sustained pace of improvement in child survival. Though unheralded abroad, or even in South Africa itself, the recent drop in Coloured infant mortality rates represents a major achievement in public health.

What can account for this improvement? Certainly not trends in family formation; Coloured illegitimacy ratios are high and have been rising. By 1984 more than half of the Coloured children born in South

Africa were probably born out of wedlock.[11] The purchasing power of the Coloured population increased between 1970 and 1985 but not at a markedly faster pace than in the 1950s and 1960s, a time of minimal measured gains in infant survival.

Much of the decline in infant mortality may ultimately be explained by an increase in the availability and quality of government services (including health care) extended to the Coloured population over these years. By the early 1970s the South African government had indicated a political interest in the material well-being of the country's Coloured group. The Trent Commission inquiry into the welfare of South Africa's Coloured population, for example, was formally convened in 1973. To a greater or lesser degree, poverty in the Coloured community is today sanctioned as an area of legitimate official concern; one may suspect that the material well-being of South Africa's Coloured population has been affected significantly by this shift in government attitude.

And what of South Africa's Blacks? Given their rapid and apparently accelerating pace of natural increase, one may infer that health levels have improved in recent decades. But how much more than this can one confidently say?

In light of the limited nature of the official data collected about the Black population, our picture of health is necessarily incomplete. Even so, the picture may be surprising. Health progress has been dramatic in large pockets of the Black population. One such place, interestingly enough, is Soweto, the enormous and politically volatile Black township adjacent to Johannesburg. If local health authorities are correct, the infant mortality rate in Soweto in 1980 would have been twenty-eight per thousand. This would be almost a 90 percent drop from rates recorded three decades earlier. If these numbers are accurate, infant mortality in 1980 was in all likelihood lower in Soweto than in the Soviet Union.[12] The million or more persons in Soweto would appear to be the healthiest population in all of black Africa.

This is not to say, however, that health levels for South African Blacks as a whole are markedly superior to those of black populations in the rest of the sub-Sahara: far from it. For a variety of reasons, Soweto is an exception to the rules of Black South African health. Research by C. H. Wyndham of the South African Medical Research Council suggests that the infant mortality rate for Blacks in urban White areas in 1980 was about eighty-six per thousand—well over three times the level in Soweto.

On the basis of these figures, one may state with some confidence that contrary to a common adage South African Blacks are *not* the healthiest black population in the sub-Sahara. The U.S. Census Bureau estimates the 1980 infant mortality rate for all of Kenya (not just its urban areas) to have been sixty-nine per thousand—almost a fifth lower than Wyndham's figure for urban South African Blacks. If Census Bureau estimates are accurate, infant mortality may also be lower in Botswana and Zimbabwe than in urban Black South Africa.

Purchasing power for South Africa's urban Blacks may well be higher than for any other black population on the continent, but it has not been adequate to purchase the level of health enjoyed by other African populations with distinctly lower incomes. Government action—rather, the lack of it—may help explain this apparent paradox. As may be seen in table 7–3, per capita incomes for the Coloured and urban Black populations were roughly comparable in 1980, yet infant mortality rates were almost 50 percent higher for urban Blacks. Such disparities suggest that even those Blacks allowed to live in urban White areas are not yet viewed by government as legitimate beneficiaries of the sorts of services the state can provide.

Whatever may be said of the Black health situation in the cities, it appears to be much worse in the countryside. Mortality is particularly high on the Bantustans. Black infant mortality for rural areas as a whole cannot be estimated with any precision. Dr. Wyndham's colleagues, however, conducted an extensive health survey in Ciskei and placed its 1980 infant mortality rate at about 130 per thousand.[13] Whatever improvement over earlier levels this rate may constitute, Ciskei's infant mortality rate in 1980 would not by this estimate readily distinguish the area from the host of sub-Saharan countries whose populations are commonly said to suffer from ineffective administration, misrule, and official contempt for the well-being of local peoples.

Demographers have noted a rough correlation between infant mortality and general life expectancy. They have developed a number of different schedules and equations to trace out these different patterns, as observed in different populations. Using one of these schedules—the United Nations General Life Tables[14]—one may calculate the sorts of life expectancies that might be associated with some of the Black infant mortality rates reported for South Africa today. With this set of schedules, Soweto's 1980 infant mortality rate would correspond with a life expectancy at birth in the low seventies—about the same as for South Africa's Whites. For Blacks in urban South Africa as a whole, the 1980 infant mortality rate would suggest a life expectancy in the mid-fifties. For Ciskei the 1980 rate would be consistent with a life expectancy at birth in the mid-forties (about what would be guessed for many of the poorer sub-Saharan countries). Life expectancy, by these indications, may differ by more than a quarter of a century today between certain rural and urban areas for South African Blacks.

Nutrition

The nutritional status of a population is more difficult to assess than is commonly appreciated. Short of intensive clinical examination—hardly a feasible procedure when millions of persons would need to be tested—the evaluation of a group's nutritional well-being will necessarily be incomplete; many of the easiest, and most conventional, measures may be misleading. Various measures, however, can give insight into the prevalence and severity of malnutrition.

Insofar as nutrition has a strong effect on health, life expectancy at birth (previously discussed) may speak in a general sort of way to nutritional well-being. Death rates for certain age groups, however, are considered a more sensitive reflection of the nutritional condition of an entire population. One such group in any society is thought to be children over one and less than five years of age.

According to official South African statistics, the death rate for White children aged one to four in 1983 (the latest year for which such figures are available) was less than one per thousand—somewhat higher than for many contemporary Western societies but well below the level at which one would take malnutrition to be a deadly problem. The death rate for one-year-olds to four-year-olds in South Africa's Asian community was similar to the White rate; whereas Asian child mortality had been three times the White rate as recently as 1970, it was within 15 percent of the White rate by 1983. Child mortality for the Coloured population was about five per thousand in 1983—about the same as the White rate during the poor White problem years in the late 1930s and more than five and a half times the contemporary White rate. Even so, child mortality for the Coloured population has been declining rapidly over the past decade and a half; as recently as 1970 child mortality was more than twelve times greater for Coloureds than Whites.

Reliable estimates for child mortality do not exist for South Africa's Blacks. Dr. Wyndham and L. M. Irwig, a colleague at the South African Medical Research Council, have estimated that the child death rate for urban Blacks in 1970 was more than fifteen per thousand[15]—higher than the World Bank's estimate for child mortality in India in the 1980s. Child death rates, and the prevalence of serious malnutrition, were presumably even higher in the Bantustans.

More information on the nutritional well-being of South Africa's population may be gleaned from consumer spending patterns. These are significant insofar as they reflect an assessment of well-being and need on the part of *households themselves.* With affluence families typically choose to devote a smaller proportion of their expenditures to food and to buy tastier, higher-cost calories with their food budgets.

The University of South Africa's Bureau of Market Research has periodically surveyed South Africa's population to prepare estimates of personal consumption expenditure (PCE) by racial designation. According to their research, in the mid-1970s South African Whites devoted about 18 percent of their PCE to food (excluding alcohol and tobacco)—slightly more than in the United States or Canada but less than in the United Kingdom. For Asians the share of food in PCE was almost twice as high (32 percent)—about the same at the time as for Hong Kong or Italy. For Coloureds the share of food in personal consumption expenditures was put at 36 percent—about the same as in Colombia or Malaysia.

For South Africa's Blacks the share of food in PCE was estimated at 41 percent—just about the same as in Kenya at the time. This figure, however, was an average of two distinct and separate patterns. On the one hand urban Blacks had a slightly lower share of food in PCE than did Coloureds (35 percent); on the other hand homeland Blacks devoted fully 48 percent of their PCE to food. This would have been closer to estimated consumption patterns in impoverished Ghana than to those of comparatively developed Kenya.

In recent decades the quality of diet appears to have improved considerably in urban areas for the non-White population. According to Bureau of Market Research, surveys the share of food expenditures allocated to meat in Black Pretoria households rose from 26 percent to 38 percent between 1960 and 1985. Similar trends were in evidence in other Black, Coloured, and Asian populations surveyed in cities over those years.

Comparable trends cannot be plotted for the rural Black population because the requisite figures are not available. Some recent expenditure data would be consistent with nutritional improvement for some of the homelands, but the household food budget continues to suggest that South Africa's cities remain enclaves of nutritional privilege for those Blacks fortunate enough to reside in them.

If one were to look for a summary measure of trends in diet in South Africa as a whole, the share of food in PCE might be the best number available. According to the South African Reserve Board, food, beverages, and tobacco accounted for 34.4 percent of total personal consumption expenditures in South Africa in 1948. According to the United Nation's *National Accounts Statistics* series, the corresponding figure for 1983 was 34.5 percent. In the modern world, it is highly anomalous to witness such stasis; even in poor countries, the share of food in personal consumption has tended to decline over time. The anomaly speaks to the peculiar nature of poverty in modern South Africa. If material circumstances have been improving over time within each racial designation, the enormous (and enforced) differences between these groups, manifest most clearly in the rise of the homelands population, have been sufficient to obscure this progress in a quick glimpse at the country as a whole.

Literacy

Education plays an instrumental role in the escape from material poverty. Exceptional individuals may rise to affluence without the ability to read or to write, but entire societies cannot. Even though literacy is a slippery quantity, it is nevertheless useful to get some idea of its prevalence—or more properly of the prevalence of illiteracy—to appreciate those aspects of poverty associated with ignorance and social isolation.

The 1980 census gives information on self-assessed illiteracy for the entire population within the RSA's boundaries at that date. According to these numbers the illiteracy rate for White adults (fifteen years of age and older) was less than 1 percent—similar to self-assessed rates of illiteracy in Western Europe, Japan, and Australia. For Asians the rate was 8 percent (3 percent for men, 13 percent for women)—about the same as reported in Korea and Taiwan. For the Coloured population the illiteracy rate was about 16 percent—about the same as the figure given for Mexico. For Blacks the illiteracy rate was recorded at about 34 percent. If correct, that would be lower than the recorded illiteracy rate for any black African state and would roughly equal the rates reported for such places as Bolivia and Turkey.

Once again, however, the Black "average" conceals stark differences between urban and rural areas. In urban areas the proportion of Black adults indicating no education (a reasonable proxy for illiteracy) in 1980 was about 21 percent—about the same as Thailand or the Philippines. In rural areas, by contrast, the proportion of adults with no formal education was placed at more than 50 percent. According to the UNESCO *Statistical Yearbook,* several African countries—including Zimbabwe, Swaziland, Uganda, and Kenya—had lower measured or projected rates of illiteracy than this figure in 1980.

More than many other aspects of poverty, illiteracy is affected by government policy (schooling). According to South Africa's 1980 census about 20 percent of the Blacks fifteen to twenty-four years of age had received no formal education at all. This fact will influence Black illiteracy for years to come. As a practical matter, it will be quite unrealistic to expect Black illiteracy to decline to the current Coloured level in less than a generation. One may note, moreover, that the 1980 figure for Blacks excluded those living in independent homelands, whose educational attainment was likely to be particularly low. This less-discussed facet of educational policy may account for as much as a third of the improvement in Black literacy in the RSA between 1970 and 1980.

While public spending does not always indicate the level of official concern with a problem or reflect the quality of services provided, the differences in public educational expenditure in South Africa are nonetheless informative. In 1985 educational expenditure per pupil (primary and secondary school) was roughly twice as high for Whites as for Asians. Per capita expenditure for Coloured pupils was roughly a third of the White level. For the Black child per capita expenditure in White areas was somewhat less than a sixth of that for Whites. And per capita expenditure for Black children in White areas was more than twice as high as for those on Bantustans within the RSA. Between 1975 and 1985 the differences in per capita educational expenditure on White and non-White pupils narrowed considerably in the White areas. What has happened in the homelands is less clear. Like its general poverty problem, South Africa's illiteracy problem is largely a problem of and on the homelands, but precisely here the central government's knowledge of and interest in the human condition seem to be least intense.

Conclusion

South Africa's patterns of poverty are unlike any other country that considers itself part of the West. Despite the vitality and vigor of its cities, this is a land in which deurbanization has been occurring. Despite considerable economic progress in the aggregate, nutritional need, as reflected in consumer spending patterns, has seemingly not lessened over a thirty-five-year span. By many measures material conditions for two-fifths of the population (homeland Blacks) would seem to be scarcely better than those in a number of sub-Saharan countries commonly said to be impoverished and poorly governed (table 7–6).

Critics who charge that the South African government is indifferent to the well-being of its citizens do not understand the meaning of these paradoxical patterns. The modern South African state is a highly developed apparatus, capable of and deeply involved in ensuring the social welfare of intended beneficiaries. South Africa's economy is extensively regulated and strongly influenced by actions of state; between 1976 and 1985, for example, the private sector accounted for

TABLE 7–6
ESTIMATED ECONOMIC OUTPUT AND STRUCTURE IN SOUTH AFRICAN HOMELANDS AND SELECTED SUB-SAHARAN STATES, EARLY 1980s

Area (1981)	*GDP per Capita ($US)*	*Agricultural Output per Capita ($US)*	*Central Government Expenditure as % of GDP*	*Devlp. Aid as % of Central Government Expenditure*
Burkina Faso	186	20	14	49
Cameroon	836	224	21	15
CAR	292	108	23	59
Ghana	236	120	11	53
Kenya	385	106	30	25
Malawi	253	100	34	22
Mali	162	67	26	73
Senegal	396	86	32	40
Sierra Leone	323	90	26	27
Tanzania	273	118	34	39
Togo	239	75	34	25
Zambia	587	104	40	24
Zimbabwe	911	149	38	9
Unweighted average	391	106	28	35
Self-governing national states (1985 rand)	219	76	109	67
Independent homelands (1985 rand)	447	70	113	NA

NOTES: Estimates for sub-Saharan states reflect numbers derived from the computation process and are not meant to suggest implicitly levels of accuracy or margins of error. Development aid figures are for total gross receipts for 1982 and central government expenditures for 1981; for South Africa, official payments and government expenditures are both for 1985–1986.
NA = not available.
SOURCES: For sub-Saharan states: Derived from World Bank, *World Tables,* 3rd ed., vol. 1 (Baltimore: Johns Hopkins University Press, 1983), and *World Development Report 1984* (New York: Oxford University Press, 1984); and Organization for Economic Cooperation and Development, *Geographical Distribution of Financial Flows to Developing Countries: Disbursements, Commitments, Economic Indicators 1982–85* (Paris: OECD, 1987). For South Africa: Derived from Development Bank of Southern Africa (unpublished figures) and Carole Cooper et al., *Race Relations Survey 1985* (Johannesburg: South African Institute of Race Relations, 1986).

barely a quarter of South Africa's net domestic capital formation.[16] Central government expenditures on social services and welfare are, in relation to gross domestic product, at roughly the same level today in South Africa as in Switzerland, Canada, and Germany.[17]

South Africa's political leaders are well aware that the state they have helped to construct is a powerful tool for alleviating poverty in whatever population it targets with benefits. The government's social welfare policies, after all, were principally responsible for the eradication of White poverty. Although Asians and Coloureds are not entitled to nearly the benefits guaranteed today to Whites, the gradual extension of social welfare spending to these groups has played a major part in their recent progress against material poverty—witness the dramatic improvement in the health of Coloured newborns in the 1980s despite the country's unexceptional economic growth.

One should make no mistake: South Africa is a welfare state. The South African welfare state, however, attends to poverty not on the basis of individual need but instead according to the entitlement of one's group. Benefits are administered, and income is redistributed, by a criterion that is misleadingly called "race"; a more accurate word might be "nationality." South Africa is a land of many ethnicities or "nations." But South Africa is to a large degree governed by and for but one of these nationalities alone: the Afrikaner nation.

The limitations of this form of rule were cogently predicted long before apartheid was even conceived. Writing in 1862, Lord Acton, the great historian and philosopher, warned Europe's nineteenth century "nationalists" that

> by making the nation and the state commensurate with each other in theory, [nationalism] reduces practically to a subject condition all other nationalities that may be within the boundary. It cannot admit them to equality with the ruling nation. . . . According, therefore, to the degree of humanity and civilization in that dominant body which claims all the rights of the community, the inferior races are exterminated, or reduced to servitude, or outlawed, or put in a condition of dependence.[18]

Lord Acton's words should be remembered in South Africa today, especially in contemplation of the situation in the Bantustans. By such economic statistics as can be assembled, output per person appears to be no higher on the Bantustans than in many other parts of Africa; agricultural output per person appears to be distinctly lower. The proportion of central government spending to domestic output, and of "foreign aid" (as Pretoria would have it) to central government spending, appears to be markedly higher than in the rest of the sub-Sahara. Although a full range of statistics is not available by which to illuminate their plight, the populations on South African Bantustans appear to live in dire poverty and in a state of extreme dependence upon charity.

Pretoria's response to the plight of the Bantustans has not been to uplift them; rather, as Lord Acton might have put it, Pretoria has outlawed them. To be sure, the Republic of South Africa has spent, and continues to spend, considerable sums in its grants to the self-governing national states and the independent homelands. But Grand Apartheid, the strategy originally enunciated by Prime Minister Verwoerd in the 1950s and only finally disavowed in 1991, postulated the eventual separation of the Bantustans from the Republic of South Africa. In this vision, the solution to South Africa's domestic poverty problem is to make foreigners of most of the poor. To date, more than 6 million former South Africans have been affected by this "solution." Despite the demise of apartheid, this disincorporation has yet to be redressed or disavowed.

South Africa's well-wishers can view the continuing disincorporation of the rural homelands from the republic only with apprehension and dismay. From the humanitarian standpoint, it augurs ill for both the populations expelled and those remaining within the republic's amended boundaries. It seems particularly ironic that South Africa's current antipoverty strategy should be so unobjectionable within the Afrikaner nation—a community, after all, imbued with the Christian ethos of universalism and so recently familiar from experience with the travails of pauperism and rural poverty. In the final analysis, poverty is necessarily a condition afflicting, and manifest in, individual human beings. Whatever its particular successes in ameliorating material need—and some of these have been remarkable—South Africa cannot come to a general solution to its poverty problem unless it comes to terms with this fact.

8

Another Look at the World Food Problem

The postwar era has witnessed profound changes in the ways mankind feeds itself. Numerous misconceptions, however, color current thinking about the world food situation. These misconceptions affect economic, agricultural, and nutritional policies of nations, and they may have a particularly important impact on the policies that are designed to relieve the distress and vulnerability of malnourished people in low-income nations. This chapter aims to correct four of the most common, and injurious, misconceptions about global nutrition and the world food economy: (1) limitations of data, (2) dimensions of the malnutrition problem, (3) the state of the world food system, and (4) the reliability of that system.

Data Limitations

Most analyses of the current world food situation are built on the assumption that the conventional data on global agricultural and nutritional conditions are reliable and accurate. It is not unusual for agricultural analysts to use these data in highly exacting calculations—as if it were possible to trace precisely changes in the production and availability of food in all areas of the world. This confidence, however, is unwarranted. Data on the world food and nutrition situation are much less reliable than many statistical agencies have encouraged researchers and policy makers to believe.

It is instructive to examine some of the recent agricultural and nutritional figures produced by the United Nations Food and Agricultural Organization, one of the principal agencies responsible for collecting, processing, and evaluating data. The FAO reports, for example, that Chad's per capita food supply rose exactly six calories per day (that is, 0.3 percent) between 1977 and 1980, and that per capita food supplies in Afghanistan and Chad differed by exactly seven calories per day (or 0.4 percent) in 1980.[1] For the periods in question, however, it is thought that upward of 90 percent of the populace of both countries was probably rural and illiterate,[2] and as much as half of the production of goods and services in both countries may have occurred in the nonmonetized economy.[3] (Even these estimates are only speculation, since a comprehensive economic survey has never been attempted in either nation; at the time of these FAO estimates, in fact, neither country had ever published a census of its population.) To make matters worse, both nations were in chaos in 1980, convulsed by the turmoil of foreign invasion and civil war. No authority in the world could hope to measure a change in national diet of scarcely a pound of grain per person per year under these conditions, as the FAO was claiming to do.

The tendency to dignify assumptions about the global food or nutrition situation with undeserved decimal places extends far beyond the borders of Chad and Afghanistan. The FAO produces an annual estimate of world food grain production that gives global grain output down to the last thousand tons[4]—implying that variations in the world harvest can be tracked with an accuracy of 0.0001 percent. The seeming specificity of these estimates is in contrast with the 15-million-ton discrepancy between FAO and the U.S. Department of Agriculture (USDA) estimates of the volume of food grains that developed countries export for the 1970s.[5] The food export system of developed countries, one may note, is perhaps the *easiest* facet of the world food economy to observe and to measure.

Estimates concerning the global nutrition situation are plagued by the same sort of false precision. FAO numbers suggest that 34 million more people were malnourished in non-Communist developing countries in 1972–1974 than in 1969–1971.[6] In presenting such estimates, the FAO implicitly suggests that it is possible not only to estimate the number of people in the third world who can be described as malnourished but also to estimate their numbers down to the last million. Yet the U.S. Bureau of the Census, widely regarded as the most sophisticated and effective enumerative apparatus in the world, estimates that it missed more than 5 million people in both its 1970 and 1980 counts of the U.S. population, that is, about 2½ percent of the total.[7] Since the FAO places the number of malnourished people in non-Communist developing countries at more than 400 million, it presumes a margin of error on its estimates of less than 0.25 percent. This would mean that its count of the numbers of malnourished people in the third world was more accurate by far than the U.S. Census Bureau's count of U.S. citizens.

Intraorganizational disputes, interagency competition, the politics of funding, and media curiosity may all reward institutions for releasing seemingly precise and scientific statistics on the world food and nutrition situation. The accuracy of such figures, however, is limited by the capacities of institutions in rich and poor nations, by the peculiarities of agricultural and nutritional phenomena, and by the nature of poverty itself.

A generation or more after independence, many developing countries have only a limited ability to collect and to process even basic crop data. (Statistics from the centrally planned economies moreover present special difficulties that are seldom considered in the course of analyses of the global food economy.) Nutritional surveys are tenuous exercises under the best of circumstances. The Health and Nutrition Examination Survey (HANES) study conducted in the United States between 1971 and 1975, for example, indicated that carefully derived and widely accepted estimates of caloric consumption for certain age groups in the American population may have been 20 to 30 percent too high[8]—and it is much easier to measure food consumption in the United States than in low-income societies. Most important, there is an inverse relationship between poverty and knowledge. In every society, precisely those groups that are most in need are the ones most difficult to identify, to reach, and to help through official institutions. The poor generate less statistical information about their lives and needs than do more affluent groups. Governments find it extremely difficult to learn about citizens who are socially invisible, geographically remote, or politically without a voice. Furthermore, governments and international institutions have the least reliable information about the aspect of the international food situation that evokes the greatest humanitarian concern: serious undernutrition.

Dimensions of the Malnutrition Problem

According to reports by organizations that monitor global trends in agriculture and nutrition, undernutrition is extremely common—almost typical—in developing countries and may be worsening in most of the world's poorer regions. In 1974 the FAO stated that as of 1970, 434 million persons in non-Communist developing countries—or about a quarter of the estimated population of those countries at that time—were surviving on diets at or below the semistarvation level.[9]

Three years later the International Food Policy Research Institute (IFPRI), the social science arm of the Consultative Group on International Agricultural Research (CGIAR), concluded that as of 1975, between 1.19 and 1.31 billion persons in non-Communist developing countries—or about three-fifths of their estimated 1975 populations—were "too poor to have an adequate diet, based on caloric intakes."[10] And in 1980 the U.S. Presidential Commission on World Hunger declared that "the world is even farther" from the goal of relieving malnutrition than it had been in 1974 and warned of "growing . . . concern over the continuing deterioration of the world food situation."[11]

It might reasonably be assumed that such definitive judgments by such influential authorities on so important a topic would be supported by careful and well-documented studies. This does not appear to be the case. Each of these pronouncements appears to have been made on the basis of extremely limited data, many of which were of obviously questionable reliability. In those instances where a methodology for estimating the extent of hunger was outlined (the 1980 presidential commission never revealed the reasoning behind its assessment), it is evident that the approaches adopted would be incapable of accurately representing the extent of malnutrition *under any circumstances*, even if perfect data had been available.

The FAO based its 1974 assessment of malnutrition in the developing countries on its food balance sheets, which attempt to estimate per capita availability of calories, protein, and other nutrients within a country. But for most of the countries for which the FAO made such computations, basic figures on national population and aggregate food availability were then, and remain today, a matter of some uncertainty.

Accurate and uninterrupted tracking of population totals requires essentially complete registration of births and deaths. Near complete registration of vital events currently takes place in about forty-five countries in Asia, Africa, and Latin America, but these account for less than a tenth of the total population of the developing nations.[12] An even smaller fraction of the hunger problem in developing countries is accounted for by these comparatively well-covered countries, since comprehensive social statistics can be compiled only by relatively developed societies.

Population data are less certain in precisely those places where undernutrition is likely to be more severe. All but a handful of developing countries have by now conducted at least one census, but these returns do not always serve as reliable indicators of current population. India's 1961 census was carefully planned and effectively implemented, but by 1971 projections based on it were off by 25 million people—almost 5 percent. In areas where a national census has never been conducted, margins of error on population estimates are necessarily higher. In 1970—the year the FAO focused upon in its 1974 assessment of global malnutrition—twenty-six countries in Asia, Africa, and Latin America had never held a national census;[13] their total populations for that year may have approached 300 million. It is hard to imagine how a serious food balance sheet could be produced for any of those societies.

Counting people is a comparatively easy task next to estimating a country's total food supply. In regions where statistical capabilities are limited, government infrastructure is weak, or any substantial portion of total food production circumvents monetized marketplaces, the accuracy of estimates of aggregate food availability cannot be great. Yet these conditions characterize many of today's developing societies—and all of those in which undernutrition is thought to be particularly prevalent. India's crop survey system, for example, is considerably more accurate than those of many other low-income nations, yet estimates of grain output are still tempered by margins of error of at least 5 percent,[14] and these crop surveys cannot even attempt to make systematic estimates of output for starchy root crops (like potatoes, yams, and cassava), which tend to be produced for home consumption. As it happens, there are no reliable estimates of output for starchy root crops for Latin America, Asia, or Africa.[15] Yet such foods of poverty account for a substantial fraction of total food supplies for many disadvantaged populations in these areas.

It would appear that, as a result of practical problems, the margins of error in the FAO's food balance sheets frequently exceed the differences the sheets were constructed to measure. But there is an even more fundamental problem in attempting to divine the extent of undernutrition from a food balance sheet. Undernutrition is determined not by national averages but by the circumstances of *specific individuals* within a country. An estimate of average caloric availability cannot be converted into an estimate of availability for individuals without some overriding assumptions about the distribution of food in the society in question. The FAO has apparently never clarified the method used in its 1974 study of converting average food availability estimates into estimates of the extent of nutritional shortfalls for specific individuals.

IFPRI took a different approach to estimating the incidence of malnutrition in developing countries. It began by estimating the distribution of incomes in different regions; then it estimated the relationship between calories and income in these regions; finally it computed the fraction of those various national populations that would be predicted, on the basis of income distribution and income-calorie relationships, to be consuming less than the average caloric recommendation that had been set by a joint FAO–World Health Organization (WHO) working group.

In practice, each stage of this exercise was complicated by serious problems. National income estimates are much less reliable than is commonly appreciated, even for developed countries. Questions of standardizing measurement and valuation are answered less satisfactorily. (According to a World Bank–sponsored project, adjustments to account for purchasing power would triple India's nominal per capita GDP.)[16] And estimates of income distribution are even more error-prone than estimates of national income. World Bank and Organization for Economic Cooperation and Development (OECD) estimates of the income share accruing to the poorest 40 percent of France's households, for example, have differed by as much as half.[17] Few developing countries would seem to be in a position to monitor their citizens' income as closely as France does.

Estimates of the relationship between income and calories in developing countries are not much better. As a rule rich people eat better than poor people, but in most developing countries the national data are not sufficiently precise to bring statistical meaning to this broad generalization. In Indonesia—a developing nation whose consumption patterns are more extensively surveyed than many others—one can currently choose among surveys whose average estimate of per capita intake differ by as much as 400 calories per day; not surprisingly the line one plots between income and food intake depends significantly on the data sample one selects. There are reasons to expect both national data and quick sample surveys to misrepresent consumption patterns in developing nations, particularly for the poor; these include the virtually irresolvable problem of locating representative samples of households at the lowest spectrum of the income scale, the difficulties of determining the actual situation and condition of interviewees, and the technical question of divergence between short-run and long-run caloric adjustments to changes in income (a problem made more salient by seasonal variations in food output, temporary fluctuations in income, and broad patterns of national income growth).

There is little reason to think that the distribution of food intake among income classes could be calculated with much accuracy for more than a handful of developing countries—and then only for developing countries for which undernutrition is least likely to be a problem. Even if such distributions could be generally computed, they would not in themselves reveal the incidence of malnutrition in poor societies. Malnutrition results from the gap between intake and need; it cannot be measured by intake alone.

There is considerable controversy over the precise level of average caloric requirements for the different developing countries. New studies suggest that FAO/WHO minimum recommended dietary allowances may provide 10–20 percent more calories than would be necessary for an *American* reference population to maintain body weight.[18] Since the principal nutritional problem in the United States is obesity, FAO/WHO standards may overstate requirements for developing countries by an even more substantial margin. By itself, an overestimate of average requirements could seriously affect estimates of the incidence of malnutrition. But an even more dubious premise is built into the IFPRI calculations. IFPRI numbers assume that anyone receiving less than the stipulated average requirement is automatically undernourished. This ignores the fact that nutritional needs within any population vary. Indeed half of any population by definition needs less than the "average" requirement. By the IFPRI logic, a country in which everyone obtained exactly enough food could be classified as having a 50 percent incidence of undernutrition. This may explain why World Bank exercises employing the same approach have concluded that 40–50 percent of the people of Hong Kong and Taiwan live on inadequate diets[19]—although life expectancies in Hong Kong and Taiwan roughly match those in North America and exceed those of several European nations.

In the age of the problem-solving state, false precision about the world food and hunger situation poses real dangers to vulnerable groups. It increases the risk that relief and development policies will be formulated and pursued on the basis of misinformation. It complicates the evaluation of programs to help those groups most in need, and frustrates the correction of injurious aspects of such programs. And it opens the possibility for unpredictable oscillations in the current reading of the food and hunger situation, leading to erratic interventions and perhaps eventually to a reluctance of political figures to commit their reputations and resources to sustained efforts to alleviate hunger systematically.

Despite the apparent lack of reliable data on the food situation for so many developing countries, it is nevertheless possible to talk with some confidence about trends in the prevalence of the most serious *manifestations* of undernutrition in most of the world's poorer regions. An accurate and widely available functional indicator of severe physical distress within national populations is the death rate.

For some poor nations, particularly those in sub-Saharan Africa, current mortality levels are still uncertain;[20] in others mortality can be ascertained only retrospectively, through the analysis of censuses or occasional surveys. But even so, mortality rates are the single most accurate and most meaningful indicator of living conditions in devel-

TABLE 8–1
LIFE EXPECTANCIES AT BIRTH IN DEVELOPING REGIONS OR COUNTRIES, 1955–1980
(years)

Area	*1955–60*	*1975–80*	*Change over 2 Decades*	
			Years	*%*
Northern Africa	44.1	53.8	9.7	22
Latin America	54.0	62.6	8.6	16
Caribbean	55.2	62.8	7.6	14
Middle Americas	53.5	62.9	9.4	18
Tropical S. America	52.7	61.4	8.7	17
Temperate S. America	62.0	68.5	6.5	11
Southeastern Asia	43.9	54.2	10.3	24
South Asia	43.1	52.7	9.6	22
Southern Asia	42.4	51.4	9.0	21
Western Asia	47.7	59.4	11.7	25
East Asia	46.6	66.6	20.0	43
Hong Kong	63.2	72.1	8.9	14
Korea	52.6	65.6	13.0	25

NOTE: Decimal points as printed in source.
SOURCE: United Nations, *World Population Prospects: Estimates and Projections as Assessed in 1984,* Population Studies 98 (New York: UN, 1986).

oping countries of any that are currently available.

Since the end of World War II, death rates for all surveyed populations in developing regions have fallen markedly. Indeed the precipitous nature of the postwar mortality decline in developing nations is acknowledged in the term "population explosion": falling death rates, not rising birth rates, were the main cause of the acceleration in natural population increase in poor countries.

Recent UN estimates illustrate the broad postwar changes in mortality in the developing regions. (Figures for sub-Saharan Africa are not reproduced here since those estimates rest principally on supposition.) Table 8–1 shows estimates (arguably, overly precise) for life expectancy at birth for different regions of the developing world. In every region for which reliable estimates are made, life expectancy rose dramatically between the late 1950s and the late 1970s. Despite variations between regions and countries, improvements in life expectancy were broad-based and substantial, with regional increases typically approaching a decade for this twenty-year period alone. Although life spans generally grew at a slower pace in the healthier nations of

TABLE 8–2
ESTIMATED CHANGES IN INFANT MORTALITY RATES IN DEVELOPING COUNTRIES, 1955–1980
(per 1,000 live births)

			Change over 2 Decades	
Area	*1955–60*	*1975–80*	Difference	%
Northern Africa	176	117	59	34
Latin America	112	70	42	38
Caribbean	110	72	38	35
Middle Americas	107	65	42	39
Tropical S. America	120	77	43	36
Temperate S. America	80	43	37	46
Southeastern Asia	135	84	51	38
South Asia	163	115	48	29
Southern Asia	172	128	44	26
Western Asia	179	99	80	45
East Asia	167	39	128	77
Hong Kong	54	13	41	76
Korea	100	35	65	65

SOURCE: United Nations, *World Population Prospects: Estimates and Projections as Assessed in 1984,* Population Studies 98 (New York: UN, 1986).

the developing world, no regions appear to have experienced a rise of less than 10 percent, and for poorer areas the increase was more typically near 25 percent over this twenty-year period.

A principal factor depressing life expectancy in developing countries is the high death rate for infants and children: World Bank studies suggest that as much as two-thirds of the difference in longevity between developed and developing countries can be traced to differences in survival rates for children under five.[21] Table 8–2 presents UN estimates of changes in infant mortality in different developing regions. For those countries for which data are reliable, significant drops in infant mortality appear to have taken place between the late 1950s and the late 1970s.

Sub-Saharan Africa's mortality trends cannot be traced with confidence. Nevertheless, there is reason to believe that life expectancy has risen, and infant mortality declined, in that region as well. Despite the difficulties in measuring it, there is little doubt that population growth in sub-Saharan Africa has accelerated consequentially since the 1950s; sub-Saharan Africa is widely thought to have the highest rate of population growth of any major region in the world. Only a portion of that acceleration is likely to have been caused by increases in fertility (and increases in fertility, insofar as they have occurred, may also speak to improvements in health). Moreover, such censuses as can be relied upon in sub-Saharan Africa indicate a pronounced differential between life expectancies in urban and rural areas.[22] Most African countries are thought to be experiencing rapid urbanization. Other things being equal, this would suggest that life expectancy would be rising.

Mortality is not a perfect measure of nutritional change. Improved nutrition is only one of a number of forces that have been pushing death rates down in developing countries: others include improvements in education, the upgrading of hygiene and sanitation, the extension of mass health services, medical innovations, improvements in communications and transportation, and in some areas improvements in civil order. Even so, the extent to which improvements in nutrition—both direct and indirect—have powered the reduction in mortality in developing countries has frequently been underestimated in the past. Sri Lanka's abrupt jump in life expectancy in the first few years after World War II, for example, was long described as a technical fix: a triumph of DDT over the anopheles mosquito. Only years later did researchers realize that abrupt and rapid drops in mortality had also taken place in Sri Lanka's highlands—where malaria had never been a problem. In both highland and lowland regions, improved access to food for broad segments of the population had been a key to the health explosion.

Technical improvements that depress death rates often do so by improving nutrition *indirectly*. Because of the close connection between disease and malabsorption of nutrients, many public health measures result in the improved feeding of a population, even if caloric intakes do not rise: curing a case of severe diarrhea can save a child as much as 500 calories a day.[23] And improvements in communications and transportation make it possible to move food rapidly and efficiently within a nation, thereby reducing the risk of famine at any given level of per capita food availability.[24] Social and technological advances mean it is possible to feed a population today better with the same amounts of food per person than could be done a generation ago. Neither food balance sheets nor food distribution estimates can demonstrate this essential change despite its evident importance to the lives of the poor.

If high and unstable death rates can be taken as a proxy for a high social risk of life-threatening undernutrition, it can be seen that worsening hunger is an anomaly peculiar to a few developing countries, not the rule governing the whole. Though different in most other ways, these few unfortunate areas have had similar political experiences. These societies that have as a whole been gripped by increasing instability in death rates have usually had governments that actively prevented their citizens from feeding themselves: by disrupting civil life and normal market operations (Kampuchea, East Timor) and by preventing relief operations from going into action (Uganda, Ethiopia). (China experienced both after the Great Leap Forward.) When specific regions or peoples within a country have suffered from fluctuations in mortality rates *not* attendant on government design, it has generally been because they were socially and geographically isolated (like the nomads on the fringes of the Sahel or the pygmies in Central Africa): beyond the reach of either the markets or the relief bureaucracies that might otherwise have provided life-saving food and supplies. Such seemingly "natural" disasters, however, have cost vastly fewer lives than the political famines of the past generation.

Variability and Instability in the World Food System

In recent years a number of influential reports have suggested that the world food system is threatened by a tendency toward diminishing stability. A major econometric study at IFPRI concluded that variability in cereal production, measured in terms of annual deviations from trend, has been rising in the world as a whole since the early 1960s. On the basis of that analysis, IFPRI researchers argued that the world food system was becoming less stable.[25] Other studies have not only pointed out a tendency toward increasing instability in food production but also have sought systematic explanations for the phenomenon—including the suggestion that heightened instability may be intrinsic to the process of agricultural modernization by which the world's growing demand for food must be met. Analyzing grain production data for India, one IFPRI economist has concluded that India's level of instability in food production has risen markedly in recent years, and that this rise was directly related to the spread of the science-based farming techniques known as the Green Revolution.[26]

While such depictions of the world food situation may have a certain superficial plausibility, closer inspection reveals that they are inadequate in two major respects. Computational exercises purporting to demonstrate increasing variability in global or regional food production are compromised by serious problems that have yet to be addressed or even acknowledged. Even more important, however, these exercises have betrayed a basic confusion about what constitutes instability in a system that is ultimately meant to serve man's needs and demands. For human beings, stability in the food system is measured not by changes in the size of the harvests but by changes in the ability of households and individuals to command access to food. These are different concepts. Variations in production in and of themselves can tell us nothing about whether the instability of a food economy is increasing or decreasing, nor can they by themselves indicate anything about the nutritional vulnerability of populations of the different regions of the earth.

The claim that variability in food production for the world as a whole, or for poorer regions of the earth, has been subject to secular increase over the past few decades appears to be supported by rigorous statistical analysis. Upon closer examination these computational exercises turn out to be compromised by a number of significant, and indeed irresolvable, problems. The first problem concerns the data themselves. The reliability of current estimates of global cereal output (cereal output, not total food production, is typically measured for variability in these exercises) leaves much to be desired, as noted. These estimates are even less reliable for poorer regions of the earth. And all estimates are less reliable the further one probes into the past. Unfortunately one cannot divine a trend in production variability without relying heavily on data from the past. In a sense the data and the statistical procedures that must be used in analyzing them are in conflict: the most reliable estimates are insufficient in number to form a base from which statistically meaningful generalizations about trends in variability may be made, and yet any sample large enough to provide a statistician with an acceptable confidence level must in reality depend on data whose own margin of error may well be greater than the annual changes purportedly being plotted by computations of variability.

A second problem concerns the method by which variability estimates are constructed. When variability is defined as deviation from trends, as it is in these exercises, it is exceedingly important to identify and specify trends properly. Should one take two periods—the first with an accelerating trend in food production, the second with a decelerating trend in food production—and treat them as if they were a single period, a much higher estimate of variability will result than was actually experienced in either period. Estimates of variability in food grain output then are highly sensitive to the researcher's choice of endpoints. One's conclusions about trends in variability can be vitiated, or even reversed, by a seemingly slight change in the dating of the periods under examination.

A third problem complicates the interpretation of these calculations: it involves the interplay between preferred, high-cost cereals (like wheat and rice) and "inferior," low-cost root crops (like cassava and

FIGURE 8–1
PRODUCTION SHORTFALLS AND PRICE CHANGES IN THE GLOBAL GRAINS ECONOMY, 1961–1991

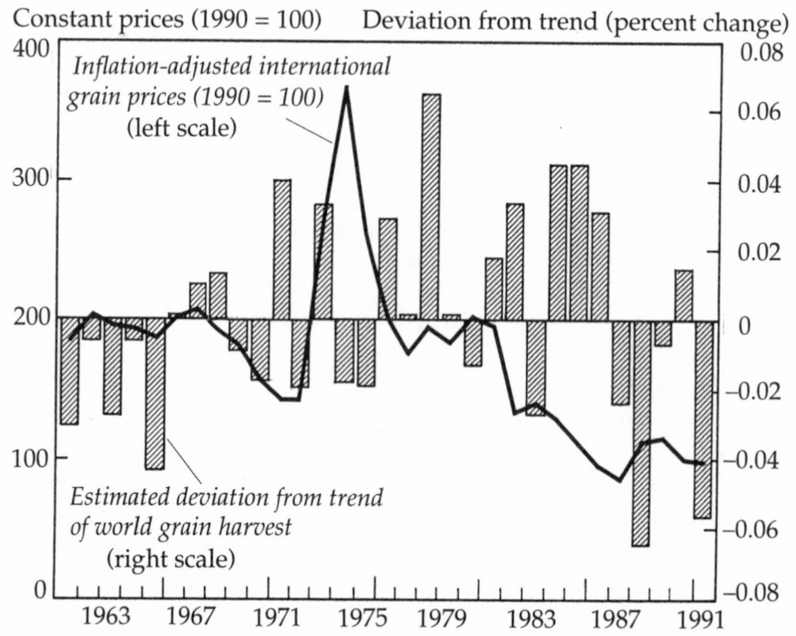

SOURCES: U.S. Department of Agriculture and the World Bank.

yams) in world production. Generally speaking, a rise in income leads consumers to reduce the proportion of inferior foodstuffs in their diet; conversely a drop in income generally results in increased reliance on low-cost foods. Because of such substitutions, production of preferred and inferior foodstuffs tends to be inversely correlated. But there are currently no reliable estimates for the production of these inferior foodstuffs, either globally or for the world's poorer regions. This can mean only that variations in actual caloric availability are exaggerated for the world as a whole by the proxy measure of cereal output. Exaggerations tend to be greater for poorer countries. And they are more pronounced in countries whose dietary patterns are beginning to be changed by affluence.

While estimates of global variability in food output must be treated with caution, we can be far more confident about variations in the international market price of foodstuffs. Such prices are not only comparatively easy to measure, but they provide meaningful information about the scarcity of these commodities. Figure 8–1 compares one set of estimated deviations from trend in the global cereal harvest with inflation-adjusted, dollar-denominated changes in the international market price of a composite basket of food grains (including soya). There is no evident correlation between designated surpluses or shortfalls—in the sense of deviation from trend—and changes in the perception of scarcity as expressed by international markets. Interestingly enough, the correlation is not meaningfully improved by choosing different endpoints for the computation of trend, using different institutions' estimates of global cereal output, or changing the weighting of the cereals within the composite price basket.

If production variability were a meaningful measure of instability experienced by food grain purchasers, we would have expected the shortfalls and surpluses in figure 8–1 to correspond, perhaps even to correspond closely, with changes in international market prices. The evident lack of correspondence between production deviations and price changes in that chart underscores the difference between variability and instability in the world food system. Variability in this context is a *physical* concept, representing the change in the mass of a harvest from one year or season to another. Instability by contrast is an *economic* concept. It describes the prospect for equilibrium between the forces of supply and demand.

It is impossible to determine whether the economic stability of a food system is increasing or decreasing solely by examining changes in production. Prospects for equilibrium are determined by a complex process, involving supply (food production and carryover stocks), demand (as articulated through the incomes of consuming households and the purchases of businesses and public institutions), and the adjustment mechanisms that intermediate between them.

Since variability and instability are in principle independent of one another, it is quite possible to imagine a situation in which the variability of food output increased even as economic instability in the food system diminished. An economic system with increasingly flexible prices, increasingly price-responsive stockholding strategies, and rising real interest rates could satisfy both requirements. (Precisely these conditions, one may note, characterized the international economy in the early 1980s.) Conversely, diminishing production variability could be consistent with escalating economic instability in a food system dominated by price rigidities and extramarket controls. (Those conditions are also far from hypothetical: cartelistic practices of agricultural marketing boards in both rich and poor countries have often worked to keep production fixed, allowing the purchasing power of producers and consumers to oscillate instead.)

For people, whether they are producers or consumers of foodstuffs, the meaningful measure of economic instability in their food system is their purchasing power over nutritional goods: their nominal incomes deflated by the price levels of the foods they need, or want, to consume. Surprising though it may seem, the variability of food production ordinarily plays a comparatively small part in the determination of short-term changes in a national population's nutritional purchasing power—even in nations that are quite poor. Agriculture constitutes a relatively small, and typically diminishing, fraction of national economic output in the modern world.

National income figures must be treated with caution—all the more so for poor nations. Even so, the World Bank currently estimates that agricultural output accounts for less than one-eighth of the gross domestic product for the approximately 500 million people who live in countries classified as upper-middle-income economies. For the more than 600 million people who live in countries described as lower-middle-income economies, agriculture is said to account for less than one-quarter of national output. Even for the more than 2 billion people in the World Bank's poorest category of countries—the ones classified as low-income economies—the share of agriculture in total gross domestic product is placed at well under half.[27]

To be sure, the fraction of agriculture in the total output in poorer nations is often artificially diminished, whether by prejudicial government pricing practices or by economic policies that discriminate more generally against farmers and the rural sector. Yet even in the world's poorer societies, the vagaries of the harvest no longer seem to constitute the principal source of instability in the population's short-term capability to obtain and to purchase food.

Over the long run, nutritional purchasing power is determined by changes in real income and the inflation-adjusted price of foodstuffs. For all the uncertainties inherent in statistics pertaining to changes in real income, or to changes in inflation-adjusted food prices, it is possible to speak of global trends for both. In the postwar period the *general* trend among nations has been toward increasing aggregate real per capita incomes; over the same period the *general* trend in international markets has been toward decreasing inflation-adjusted food prices. (Any regional and temporal exceptions to these tendencies have been precisely that: exceptions.) These trends suggest that, from the standpoint of consumers, the stability of the world food system has generally been increasing over the postwar period.

This is true even for the population of one of the world's largest and poorest countries. In India, per capita income is estimated by the World Bank to have increased more than half between the early 1950s and early 1980s.[28] Over the same period the real price of wheat in the international market fell more than 40 percent; that of rice, more than

30 percent; and that of corn, more than 60 percent.[29] While choices of dates, and methods of calculation, could alter these specific figures, they would not mitigate the conclusion that nutritional security has increased, not decreased, in India during the recent decades of agricultural and economic modernizations. That conclusion, moreover, is fundamentally consistent with those demographic estimates that indicate that life expectancy in India may have risen twenty years between the early 1950s and the late 1970s.[30]

Reliability of the World Food System

In journalistic, academic, and governmental circles the view remains current that the vulnerability of given nations, and even of entire regions, to sudden and unexpected shocks in the food system has increased since the late 1960s. The perception that the global food system has been growing more susceptible to dangerous dislocations seems to have been encouraged by some of the recent changes in the structure and operation of world food markets. An increasing fraction of the grain consumed in less-developed countries, for example, is being purchased from abroad. Furthermore, these nations must buy their imports from a marketplace in which neither short- nor long-run prices can be predicted with any certainty, but within which prices fluctuate constantly. Binding international treaties to secure the price and quantity of grain available to specific nations are generally absent from the international food trade today—in contrast to some periods in the past. Finally, major participants in the international cereal trade have been changing the rules by which they play. In the early 1980s, for example, the United States deliberately and radically diminished governmental grain holdings and gave as explicit rationale for that policy the hope that depleting these public reserves might prompt the price of foodstuffs on the world market to rise.

Each of the previous observations about the world food system is accurate. The picture they present of the working of the world food system, however, is incomplete and seriously misleading. Specifically missing from this tableau is an understanding that the world food system is a world food *economy,* built upon and animated by forces that are principally economic in derivation and expression.

True, the net import of food grains by less-developed countries has increased dramatically in recent decades. The World Bank estimates that net food grain imports for what it categorizes as developing countries have risen from an average of about 22 million tons in the early 1960s to an average of about 78 million tons in 1984; while perhaps only an imperfect approximation, these numbers nonetheless indicate an unmistakably rapid pace of growth. But the growing use of international markets to meet local food needs in low-income countries has been prompted by economic logic. While the relative price of internationally traded food grains was declining over the past few decades, the purchasing power of the developing countries increased substantially. Over the past generation, per capita income for this diverse grouping of nations may have doubled, while economic output may have tripled. The exports of less-developed countries have moreover increased at a pace more rapid than their rate of aggregate economic growth. Thus the finance of food grains and cereal imports for developing countries has required a steadily diminishing fraction of their export revenue. In the 1980s developing countries could pay for their total food grain and cereal imports with only about 4 percent of their total export earnings. The diminishing cost of grain import bills in relation to export revenues indicates not increasing vulnerability to shocks, but precisely the opposite.

The lack of a global system of comprehensive bilateral grain agreements has not proved a decisive deterrent to expanding international trade in food grains. In 1970 something like 120 million tons of cereals and soybean products moved across international borders; by the early 1980s international purchases may have exceeded 260 million tons a year. (The rate of growth in the volume of food grains moved by international markets during these years was probably about 6 percent per annum—several times the contemporaneous rate of world population growth, and distinctly higher than conventional estimates of the rate of global economic growth.) As the volume of food grains passing through international markets increased, the sources of supply diversified. This diversification was evident in the emergence of the European Economic Community as a major (if highly subsidized) cereal exporter, and in the reinvigoration of cereal exports from such nations as Thailand, Brazil, Argentina, and South Africa. The broadening and deepening of the international grain market, other things being equal, suggest a decrease in the riskiness of relying on international food supplies for national needs. Indeed, the perception of diminished risk may have been one of the factors that stimulated this rapid expansion in the international food trade.

The prices of the many foodstuffs traded in international markets fluctuate constantly; the variations in these prices, moreover, cannot be predicted with any confidence, even in the near future. But it does not follow from these observations that the process of *price formation* within these markets is necessarily erratic or unstable. Prices reflect perceived scarcities; they may be expected to shift as the perception, or reality, of scarcity shifts. Moreover, preventing prices from changing where there is pressure on them to do so entails real economic costs to both producers and consumers. It would demonstrably increase the riskiness of relying upon markets to meet food needs and would encourage greater dislocations in the future.

As it happens, the process of price formation in the international cereal markets appears to have become more, rather than less, stable in recent years. Increased integration of the global livestock feed economy, with the hundreds of millions of tons of cereals it may absorb or release at appropriate prices into the international grain markets has provided the world with a growing buffer against dramatic price movements in the face of unexpected shocks.

Even more important, the 1970s and 1980s saw the rapid growth of an organized grain futures market. With a value of annual transactions in the 1980s of about $300 billion, the futures market is a major vehicle for ensuring against erratic price formations. Futures contracts allow trading partners to make transactions as much as eighteen months into the future at predesignated prices. The large volume of transactions enables purchasers to hedge against price changes by continually rolling over contracts to cover anticipated future requirements. At the same time, the futures market reduces the risks by producers and merchants of attempting to maintain private carryover stocks of foodstuffs. Since the futures market reflects current and anticipated supply and demand levels, the prices registered in these contracts give signals by which to assess the profitability of releasing reserves into the market, or alternatively of holding them in stock. Growth of the futures market has facilitated a substantial expansion of commercial grain reserves. Among developed countries alone, such privately held reserves have increased from less than 20 million tons in the early 1970s to an estimated 80 million tons in the early 1980s, according to the U.S. Department of Agriculture.

In itself, the rise of the grain futures market and private commercial grain stockholdings speaks to the emergence of a more orderly process of price formation in international grain markets—a process that is much more likely to manage unexpected shocks. No less significantly, the growth of the futures market and private commercial grain holdings has contributed to the displacement of government-owned reserve stocks as an adjustment mechanism in the world food economy. There is little reason to think that the eclipse of government-owned reserve stocks should destabilize the process of price formation in international grain markets. Quite the contrary: government stocks are maintained and managed in accordance with political, rather than strictly economic, criteria and thus do not necessarily operate as price-responsive entities. In important instances governmental interventions in stockholdings have considerably *destabilized* prices in food markets or increased economic risks to customers relying upon international markets to meet their food requirements.[31]

The existence and growth of grain futures markets and commercial stockholdings have been predicated upon, and in turn have stimulated the demand for, detailed and accurate information about farm, food, and business conditions in all regions of the globe. Some developments in information systems, whether technological or diplomatic, have had specific relevance to the assessment of the world food situation.[32] Greater quantities and quality of information may well hasten change in local and international food markets. But change should not be mistaken for instability. Under ordinary circumstances, improvements in information within a competitive market provide both buyers and sellers with a better base for the decisions they make. That should *reduce* instability and thereby reduce as well the risks of participating in these markets.

Because the world food system is actually a world food economy, not all developments affecting the vulnerability of countries to unexpected shocks in the food system have a direct or obvious connection to the food system itself. We have already mentioned the importance of the growth of export revenues in developing countries in protecting those nations against disturbances in international food markets; the growth of export revenues for that grouping of countries, however, has been fueled overwhelmingly by increasing trade in minerals, manufactured goods, services, and other nonagricultural factors. In the same fashion, the risks in relying upon international food markets to meet local food requirements were meaningfully reduced in the 1970s and early 1980s by the growth of international capital markets. The period since the demise of the Bretton Woods accords has seen a growth in the pool of funds available for loan around the globe at commercial terms. By the rough estimates of the International Monetary Fund, the net flow of nonconcessional capital to less-developed countries increased from $11 billion in 1970 to $64 billion in 1983; even allowing for intervening inflation, this represents a dramatic rise. The international market for short-term credit grew with particular speed during these years: including supplier credit such outstanding liabilities to less-developed countries now exceed $200 billion at any given time. With a substantial pool of short-term capital at their disposal, the prospect of being left illiquid—temporarily unable to finance participation in the food trade in the face of an unanticipated shock—no longer looms as a serious possibility for many creditworthy developing countries.

Recent developments have not eliminated the risk of disruptions or dislocations in the world food economy. Indeed, it is difficult to imagine how the possibility of such disturbances could ever be completely forestalled from a system governed by politics, business cycles, and weather. The more useful question is whether risks in relying upon international markets for local food needs have increased or decreased as a result of recent structural changes in the world food system. The increasing commercialization of international food flows, the growth of purchasing power in the world economy, and institutional and operational changes propelled by the desire for a smooth conversion of money into food, or food into money, all suggest that the ability of the world economy to mitigate unanticipated shocks, and to protect national populations against the consequences of shocks that do occur, increased meaningfully between the late 1960s and the early 1980s.

Exceptions to Global Trends

We have argued that the world food economy has experienced three general trends in recent decades: first, a broad-based decline in the nutritional vulnerability of national populations, as reflected in changing patterns of mortality; second, an increase in the nutritional purchasing power of national populations, as measured by changes in the international cost of foodstuffs and by changes in national income; and third, a diminution in economic instability within the world food economy and a decrease in the economic and nutritional risks of relying upon international markets to meet local food requirements. But there have also been important exceptions to these general tendencies. Sub-Saharan Africa has emerged as a region of special concern to both bilateral and multilateral aid institutions in part because of the widespread perception of a general deterioration in agricultural conditions and of increasing nutritional vulnerabilities for both isolated groups and entire national populations. Furthermore the ability of the international grain markets to perform reliably in the face of unanticipated shocks was seemingly belied by the tumultuous events in the years 1972–1974, when purchasing nations were confronted by rapid and at times discontinuous rises in food prices amounting to jumps of 200 percent and 300 percent in a matter of months and by interruptions in the availability of international supplies. (Those disturbances are today commonly referred to as the "world food crisis.") The situation in sub-Saharan Africa and the problems during the world food crisis would require examination under any circumstances simply on the basis of their human implications. They deserve special attention at a time when prospects for nutritional security seem generally to have been improving.

While demographic and agricultural statistics are largely unreliable for the vast expanse of sub-Saharan Africa, indirect evidence does not contradict the conventional view that widespread food problems have emerged in the region over the past twenty years. From the 1950s until the early 1970s, sub-Saharan Africa's share of international agricultural commodity markets was measured as increasing. Since then, the situation seems to have been reversed: sub-Saharan Africa was indisputably losing its share of these export markets during the mid-1970s until the 1980s. As best can be told from international trade data, export revenues for sub-Saharan Africa stagnated between the early 1970s and the early 1980s, and export earnings in some sub-Saharan countries appear to have fallen since then. Such statistical data as can be relied upon give no indication of any significant recent increases in purchasing power for the region. And while neither comprehensive nor always reliable, there have been increasing reports of both episodic and recurrent food shortages for many groups in the sub-Sahara. Since many of the populations affected derive much of their sustenance from home production, their difficulties have been taken to mean that agricultural output is failing to respond to household requirements in a broad variety of localities.

What is the problem? Many observers have spoken of environmental troubles in sub-Saharan Africa. Such problems should not be belittled. Although the evidence is far from conclusive, the climate in some areas of sub-Saharan Africa may have entered a cycle in which unfavorable conditions are more likely than they were in previous decades. And in some areas increased cultivation or overgrazing has probably contributed to the disruption of a natural balance that had previously preserved productive capacity even in periods of adverse weather. But of far greater significance have been the impact of government policies, and the tenor of politics, in actively discouraging agricultural and economic growth.

There is little disagreement among students of contemporary African affairs that the climate for private investment and economic activity in most sub-Saharan countries has deteriorated in recent years. (See chapter 9 for more details.) Internationally business confidence in many of these countries has been eroded by systematic restrictions on foreign direct private investment, discriminatory and often substantial tariffs and barriers against free private trade, extensive controls on foreign exchange, a tendency to drain or to confiscate profitably operating foreign enterprises, and a perceived reluctance of governments and government-owned corporations to honor contractual obligations. As a result of declining confidence in the business climate, direct overseas private investment is today minimal in the entire region, and some governments appear to be unable to raise credits on international commercial markets.

If the business climate is inauspicious for international entrepreneurs, it is generally even less promising for local businessmen and domestic merchants. Unlike international financiers, these local operators lack the political influence that comes with access to substantial streams of foreign capital and also lack the recourse of moving their funds to other nations if local conditions become unattractive. Government practices in many sub-Saharan countries have increased the difficulties encountered by local peoples in pursuing many private economic activities, including farming and food marketing. Government interventions in many of the countries with the most serious agricultural problems have been deliberately directed at keeping agricultural prices low, in effect taxing farmers to subsidize the living standards of urban groups. (Urban populations may typically be more affluent than rural people in sub-Saharan Africa, and they may also typically be more politically influential.) The disincentives to increasing agricultural production or improving marketing infrastructures in many sub-Saharan countries are now considerable, often as a direct result of actions undertaken by the state.

Political decay has also played a role in shaping the climate for agriculture and business in much of sub-Saharan Africa. Since the early 1970s, many sub-Saharan countries have suffered from chronic domestic instability, war, or civil violence. Food problems generally seem to have been most acute in these areas.

Not all sub-Saharan countries are thought to have experienced economic or agricultural decline over the 1970s and 1980s. Differences in performance between some sub-Saharan countries during these years are believed to have been dramatic. Table 8–3 highlights some of these differences. While the figures are illustrative, data on African agriculture are limited; they do serve, however, to underscore a number of points about recent developments in sub-Saharan Africa. First, none of the countries thought to have experienced the most severe agricultural setbacks were located in the Sahel, the region where changes in weather are said to have had their most adverse consequences on agriculture. Second, most of the countries with the worst agricultural records are contiguous to countries with some of the best records of agricultural performance: Uganda borders Kenya; Mozambique borders Malawi; and Ghana shares a border with the Ivory Coast. Third, the agricultural problem of sub-Saharan Africa's poorest performers does not appear to be one of weak growth, insufficient to meet population increases; to the contrary these states apparently saw aggregate agricultural production drop below the levels that had been established within

TABLE 8–3

ESTIMATED TOTAL OUTPUT OF AGRICULTURAL PRODUCTS AND FOOD IN SUB-SAHARAN AFRICA, 1979–1981

(1969–1971 = 100)

Country	*Total Agricultural Output*	*Total Food Output*
Highest performers		
Ivory Coast	167	184
Kenya	154	142
Malawi	145	131
Rwanda	125	143
Lowest performers		
Angola	59	92
Ghana	93	93
Mozambique	88	91
Uganda	84	101
Sub-Saharan Africa, not including Republic of South Africa	118	120

SOURCE: U.S. Department of Agriculture, *World Indices of Agricultural and Food Production, 1974–83* (Washington, D.C.: USDA, Economic Research Service Statistical Bulletin 710, July 1984).

their borders a decade earlier.

Like the current situation in sub-Saharan Africa, the "world food crisis" of the early 1970s cannot be understood without an appreciation of the destabilizing impact of government actions on the food system. By many criteria, the years 1972–1974 should not have been a period of unusual turbulence in international food markets. Although global harvest estimates, as we have emphasized, are prone to large margins of error, there is no evidence that the years of 1972, 1973, and 1974 were a time of poor crop production for the world as a whole. Such figures as have been produced uniformly suggest that the 1972–1974 period was above trend in comparison with the previous ten years. As for changes in demand, the growth of global income is not believed to have accelerated during these years; if anything, the OPEC price shocks would have slowed the growth of world purchasing power for the period as a whole. With unexceptional changes in underlying conditions of supply and demand, other factors are needed to explain the doubling and tripling of grain prices on international markets, as well as the interruption of the basic supply sources upon which these markets had regularly relied.

One of these factors was a change in the world's monetary policies. In 1971, the year before the world food crisis, the Bretton Woods arrangement for international currency exchange broke down, in large part because of a decision by the Nixon administration to finance the U.S. budget deficit through foreign dollar obligations rather than through domestic taxes. In the aftermath of the U.S. decision to cease redeeming dollar obligations for gold, a new arrangement for the international conversion of currencies evolved, based on floating rather than fixed exchange rates. Lack of confidence in the soundness of U.S. monetary policies under a floating exchange regimen led to an erosion in the value of the dollar. This created extraordinary inflationary pressures in all commodity markets in which transactions were denominated in dollars, including the grain markets.

Shortly after the first such Nixon shock, abrupt and unexpected actions by governments of developed countries helped to trigger the world food crisis. The Nixon administration had resolved to reduce the U.S. government's grain reserves—at the time by far the single largest component in worldwide stockholdings—in part because of the prohibitive cost of maintaining them at public expense. U.S. efforts to cut reserve holdings were greatly facilitated by a major Soviet grain purchase in 1972 (a decision conditioned in Moscow by an interest in keeping domestic meat supplies stable in the face of erratic local feedgrain harvests). The United States committed more than $500 million in subsidies to help secure the deal. The U.S.-Soviet arrangement was conducted in nearly total secrecy. As a result of these conditions, the Soviet government in all likelihood was able to purchase a considerably larger volume of grain than ordinary market conditions would have warranted. Moreover, since the announcement of the agreement caught international traders by surprise, the arrangement precluded a gradual adjustment of prices or an orderly logistical response to what was an unprecedented and unexpected demand on international markets.

In the period following the Soviet-American grain deal, the ability of the international grain markets to function was further compromised by new politically imposed restrictions. In 1972 the Peruvian anchovy catch failed; as a result, in 1973 demand increased for soya products, a substitute for anchovy meal in livestock feed mixes. The major world supplier of soybeans at the time was the United States. The Nixon administration feared that rising soy prices would further stimulate inflationary pressures on meat prices, and in the process threaten the viability of the domestic wage-price control program it was then administering. In September 1973 the U.S. government placed an embargo on soya exports, even voiding outstanding contracts. Though quickly rescinded, the action affected not only the international soya market but also the credibility of all international food markets in which the United States was a major participant.[33]

Disarray in international food markets was compounded by the manifestly uneconomic interventions of purchasing countries. In 1974, for example, Taiwan placed a large order for rice in the Bangkok market. The purpose of this action remains obscure, for Taiwan was generally a net exporter of rice and the indications were that the 1974 crop would be above average. As the shipments were delivered, moreover, it became obvious that little consideration had been given to the logistics of storing the purchase; local granaries could not hold the order, so the excess was stored in schoolrooms and other public facilities ill-suited to the purpose. Eventually much of the panic purchase was lost to spoilage, supplied to Taiwan's livestock and beverage industries at concessional prices, or resold on the world market at a loss. This and other purchases conditioned by a hoarding instinct not only wasted financial resources of the importing countries but had the obvious effect of making it more difficult for international grain markets to return to equilibrium.

The international food markets did, however, return to equilibrium. Despite what may have been below-trend world harvests in 1975 and 1976, and what was indisputably an acceleration of the underlying rate of international inflation, the nominal price of grain fell substantially, and the volume of grain traded on international markets increased.

The international food markets have been subject to a number of tests since the world food crisis. World grain harvests are believed to have been below trend in the years 1979 and 1980; such a deviation, moreover, would have coincided with a period of exceptionally rapid international inflation, and with the turbulence in the financial system that was attendant on the second major round of OPEC price increases. In the early 1980s the American farm sector, which dominates the export market in many agricultural commodities, came under financial stress; according to the Department of Agriculture, as many as 15 percent of all farms were in risk of bankruptcy, and a larger fraction were thought to be under pressure to restructure or to cease operations. These and other challenges have to date resulted in no new disruptions in world food markets. Although many governments continue to pursue inefficient, regressive, or disruptive farm policies domestically, few countries are currently willing to disrupt or to destabilize the international food markets in any similarly active fashion. Such restraint from the most injurious sorts of interventions into these markets accounts in large part for the fact that the world food crisis has not yet repeated itself.

Conclusions

Most of this chapter has been devoted to correcting some of the more important misconceptions about the current world food situation. To the extent that these misconceptions shape policy, they can be expected to prove harmful to diverse populations, and to have an especially severe impact on the welfare of the poor. A number of points may prove germane to the alleviation of undernutrition and to the acceleration of material development.

First, nutritional security depends primarily upon the purchasing power of households and individuals and on the efficiency of the markets through which they may articulate it. Ambitious regimes have attempted to subordinate the production and allocation of food to direct political control and have in varying degrees succeeded in displacing economic mechanisms with political ones in their local food systems. But they have yet to replace fully—or to improve upon—the nutritional guarantees which households and individuals can derive from improvements in their own purchasing power and from productively functioning markets.

Second, in any successful national journey to general nutritional security, the importance of both agriculture and agricultural policy may be expected to decline steadily. The share of agriculture in national output tends to decline as the general level of purchasing power rises. Agriculture can make only a diminishing contribution to the increase in purchasing power as the level of national affluence rises—barring expensive political distortions. By the same token, fluctuations in purchasing power will be increasingly dominated by nonagricultural factors. Thus, a nation's nutritional security comes to depend ever more on its macroeconomic policies and "business climate," and ever less on agricultural policies alone. Fiscal and monetary policies, tax policies, budget management, foreign exchange regimens, trade strategies, credit policies, price controls directly affect the nutritional security of modern countries, even when these policies have no immediate connection to farm and field.

Finally, the hand of government is likely to be increasingly prominent in situations where local food problems intensify or suddenly emerge. In the modern world any deterioration of nutritional security for any sizable population should immediately suggest a failure of those institutions that intermediate between local households and the global economy. One of the most important functions of the national government is precisely this.

It is not hard to identify instances where modern governments have acted in ways that reduced the nutritional security of the populations beneath them. Many governments embrace policies that demonstrably reduce national purchasing power or purchasing power for the poorest strata of their nations. Many governments intervene in domestic markets in ways that diminish the capacities of these markets to service customers. And in more than one case, modern governments have actively contrived to subject some portion of their citizenry to famine.

This is not to suggest that governments can make only negative contributions to nutritional security. Quite the opposite is true. With the rise of a global economic system, prudent and responsible economic policies may be expected to confer both local and international benefits. Experiences in both rich and poor countries indicate that governments can play a limited, but helpful, role in making some markets work better. And even in the best of all imaginable futures, the careful administration of relief for needy groups will be an important way to protect distressed populations against a nutritional setback for many decades to come.

9

Investment without Growth, Industrialization without Prosperity

Four decades ago, President Truman first proposed technical assistance as a form of foreign aid. Now commonly known as official development assistance, or ODA, it has become a prominent and established feature of the international economic landscape. Virtually every Western nation today maintains its own ODA program, and with the exceptions of Japan and Taiwan, virtually every country in Asia, Africa, and Latin America is currently an ODA recipient. From President Truman's modest initial request to Congress in 1950 of $45 million, bilateral and multilateral ODA commitments have expanded enormously. During the 1980s the pace of ODA commitments has averaged about $40 billion a year. Thus in the late 1980s the value of annual ODA commitments actually exceeded the dollar volume of a number of important items in the world economy—total world trade in foodgrains being but one of them.[1]

Over the years the official rationale for this international flow of concessionary resources has varied. Among donor and recipient governments alike, however, two arguments seem to have been particularly influential. During the 1950s and 1960s, proponents of ODA commonly argued that these transfers could promote economic growth by augmenting investment in physical capital and by helping to finance government industrialization programs in low-income countries. In the 1970s and 1980s, it was increasingly argued that ODA from the West should also be used to finance social welfare services. By helping

TABLE 9–1
CHANGES IN ESTIMATED LIFE EXPECTANCY AT BIRTH IN SELECTED REGIONS, 1950–1985
(years)

Area	*1950–55*	*1980–85*	*Change (years)*
More-developed regions	66	73	7
Less-developed regions	41	57	16
Africa	38	49	11
Latin America	51	64	13
Asia	41	59	18

SOURCE: United Nations, *World Demographic Estimates and Projections, 1950–2025* (New York: UN, Department of International Economic and Social Affairs, 1988).

to meet basic human needs through support of local programs for health care, education, housing, food, and the like, the argument ran, ODA not only would directly alleviate material poverty but would enhance "human" capital and thus lay the foundation for future economic development.

True to the arguments and predictions of its proponents, the era of ODA has been a time of pervasive and dramatic material progress in low-income countries. The scope of these material changes is perhaps most vividly captured in the mortality rates and survival probabilities for these regions. As table 9–1 and 9–2 indicate, health levels throughout the low-income realm have improved substantially in recent decades. The United Nations estimates that life expectancy at birth for less-developed regions rose almost two-fifths between the early 1950s and the early 1980s, and that infant mortality rates fell about half. These estimates suggest a marked reduction in the gap of life expectancy between rich and poor countries—a reduction with obvious implications for the well-being of ordinary people in the countries concerned.

But these material changes, salutary as they are, do not in themselves prove, or even necessarily suggest, that a generation of ODA policy has successfully promoted its stated objectives. The actual impact of ODA flows on the structure, output, and productivity of recipient countries is the pertinent criterion for evaluating the contribution of ODA to economic progress.

It is a striking fact that neither donor nor recipient governments seem to have made a regular practice of assessing the impact of ODA flows on economic performance in low-income countries.[2] That neglect

TABLE 9–2
CHANGES IN ESTIMATED INFANT MORTALITY RATES IN SELECTED REGIONS, 1950–1985
(per 1,000 live births)

Area	*1950–55*	*1980–85*	*Change (%)*
More-developed regions	56	16	–71
Less-developed regions	180	88	–51
Africa	191	112	–41
Latin America	125	62	–50
Asia	181	83	–54

SOURCE: United Nations, *World Demographic Estimates and Projections, 1950–2025* (New York: UN, Department of International Economic and Social Affairs, 1988).

and its implications for present policy deserve to be examined elsewhere at length. This chapter does not presume to provide the comprehensive evaluation of ODA policy that is today so conspicuously absent from the literature of economic development. The purpose here is rather to examine some data, often overlooked, that cast light upon the development process in low-income countries over the past generation.

It is hazardous to generalize about an amalgam of economies so evidently diverse as those of the contemporary third world. Nevertheless some puzzling and indeed troubling tendencies are apparent in even the most basic and summary statistics on the countries' national output. Simply put, the economies of the ODA-receiving countries on the whole appear to be characterized by pronounced and gross structural distortions. To appreciate the magnitude of these distortions is to wonder how such pervasive and generally untoward transformations would possibly have been financed.

Industry, Investment, and Personal Consumption

As anyone who has tried to work with them can attest, economic statistics on low-income countries should be used with caution. Frequently their accuracy and reliability do not warrant exacting comparisons or sophisticated computational exercises. Their inherent limits of confidence argue for simple comparisons.

One such simple but nonetheless instructive comparison may be seen in table 9–3. The table shows recent World Bank estimates for the portion of national output associated with industry and absorbed by investment throughout the world. Even such basic estimates as

TABLE 9–3
ESTIMATED INDUSTRIAL PRODUCTION AND GROSS DOMESTIC INVESTMENT FOR SELECTED COUNTRIES AND GROUPS, 1985
(percent of gross domestic product)

Countries and Groups	*Industrial Production*	*Gross Domestic Investment*
Industrial market economies	34	20
Developing economies	36	24
Low-income economies	33	27
Middle-income economies	37	23
Sub-Saharan Africa	26	13
India	27	25
Zimbabwe	45	23
South Africa	45	21
Mexico	35	21
Denmark	24	20

SOURCES: For groups, World Bank, *World Development Report 1989* (New York: Oxford University Press, 1989), pp. 148–49; for individual countries, *World Development Report 1987* (New York: Oxford University Press, 1987), pp. 208–11.

these must be handled with care. Nevertheless the World Bank aggregates reveal a consistent and peculiar pattern. Among even the poorest low-income societies, the share of industry in total output is barely lower than for the advanced industrial democracies as a whole; moreover the investment ratio for developing countries actually appears to be higher today than in rich countries.

This anomalous situation is even more striking in the comparison of World Bank estimates for specific countries or places. As far as output structure is concerned, sub-Saharan Africa currently looks more "industrialized" than Denmark. By the same measure a number of developing countries—including Zimbabwe, Botswana, the People's Republic of the Congo, and Trinidad and Tobago—would appear to be more industrialized than Japan. Zaire seems to be more industrialized than the United States; not a single country in the European Community seems to be as industrialized as Argentina.

As for the relative level of gross investment, World Bank estimates suggest this to be lower in Germany than in such countries as Togo, Nepal, Egypt, and Costa Rica. Recent estimates put the investment ratio in Rwanda well above that in Belgium; Mali's appears to be higher than France's; and Sri Lanka's, higher than Britain's. The investment ratio appears to be higher today in the People's Democratic Republic of the Congo than in Japan. No Western country is estimated to have a gross domestic investment ratio as high as that of Lesotho.

Indeed, if one had access only to figures on the share of industry and investment in national output, one might reasonably conclude that Mexico was more developed than the United States, Peru more developed than Sweden, and India more developed than Denmark. Only when names are applied to the countries in question does it become clear these patterns speak not to prosperity but to something quite different.

More than a century of research on household budgets and consumer spending patterns has documented the strong tendency for poorer households to allocate higher proportions of their expenditures to food. The same research has underscored the general tendency for poorer households to save and to invest a smaller fraction of their income. These tendencies have been documented across cultures and across time. The findings are so consistent that they might almost be described as rules of social science. The two tendencies can be explained in human terms. Poor people act to satisfy their need for food before attending to less urgent desires, and they place a high value on the current consumption upon which their lives may literally depend.

Human though they may be, these rules do not seem to apply to economic activity in much of the low-income world today. Neither the consumption nor the food sector in many locales commands the share of national output that one might expect for low-income populations. Agricultural output, for example, accounts for an almost inexplicably low portion of the product of many less-developed countries. World Bank estimates on sectoral output may illustrate the case. According to these numbers, the share of agriculture in Zimbabwe's 1985 national output was lower than it had been in Finland in 1965. As of 1987, the fractions of GDP derived from agriculture in Mexico and Ireland were roughly the same. In 1986, by the World Bank's numbers, the share of agriculture in total output was higher in Portugal than in Gabon, higher in Spain than in Peru, higher in Denmark than in Trinidad and Tobago.[3]

As for private consumption, World Bank estimates suggest its share in output is nearly as low today for the developing countries, as a group, as for the advanced welfare states of the West. These same figures suggest that the ratio of private consumption to national output is lower today in Malaysia than in Sweden, lower in Cameroon than in Belgium, lower in Panama than in Japan, lower in Costa Rica than in the United States, lower in Kenya than in Switzerland, and lower in Ivory Coast than in Greece. By the World Bank's estimates, consumption as a fraction of gross domestic product is distinctly higher today in the industrial market economies of the West than for the developing countries as a group. International Monetary Fund estimates reveal a similar pattern.[4] According to their numbers, which trace the trends back to 1950, the share of consumption in national output has been lower for the developing countries than for the industrial countries virtually every year since 1972.[5]

Broadly speaking, neither the current patterns of output in the less-developed countries nor the incremental changes over the past generation that produced those patterns reflect the preferences that might ordinarily be expected of low-income people. One would hardly expect poor populations, of their own volition, to devote so little of their purchasing power to food and agriculture or so much to industrial produce—especially heavy, nonconsumer goods. One would hardly anticipate that poor people, for whom the perils of privation are real, would of their own accord opt to consume as low a fraction of their gross national product as the affluent West, or to forgo consumption (through the agency of gross investment) of an even greater portion of their national income than in the West.

These patterns, instead, would seem to suggest that a vast and involuntary transformation of economies has occurred in low-income countries around the world. To judge by their output, the economies of today's developing countries would seem strangely insensitive to the needs and desires of their populations. Insofar as these peculiar patterns appear to be relatively recent, it is correspondingly apparent that the developing economies on the whole have become less oriented toward articulating popular demand, and less effective in satisfying self-assessed human needs, than in earlier decades. In that fundamental sense, the distortions impressed upon low-income economies over the past generation may fairly be described as inhuman.

Consider what these distortions portend for the well-being of local populations. Imagine an economy in which the level of consumption is artificially forced to an involuntarily low level. Imagine further that within the already deliberately restricted sphere of private consumption, agricultural produce should then be diminished still further. The human consequences of such interventions should be self-evident. Other things being equal, the affected population can be expected to experience increased material privation. The incidence of material poverty, by almost any meaningful definition, would be expected to rise. And the repercussions of such interventions could be expected to be most severe precisely among the most vulnerable groups within the population. The lower the population's initial income level, the more stark would be the human consequences from such interventions.

For all the adverse human implications of artificially depressing consumption levels in low-income countries, and artificially augmenting industrial production, the prevalence of such distortions among developing countries may in at least one sense be described as a success. Reshaping national output to a more "modern" structure, after all, was an explicit objective of almost all of the many industrialization plans embraced by governments in low-income countries in the 1950s and 1960s. To the extent that these distortions are the result of industrialization policy, they may be seen as the accomplishment of deliberate government action.

In the late 1960s and early 1970s, various development specialists noted that the incidence of poverty in a number of low-income countries had remained surprisingly high despite rapid rates of industrial growth. It was at this point that the policy of "basic human needs" was proposed, and gradually adopted by low-income governments.

Coming as it did immediately after the vogue of forced-pace industrialization, the advent of a policy of basic human needs has been quite fascinating. Ordinarily the simplest method of attending to the adverse consequences of an induced distortion is to cease exerting the distorting influence. Rather than relieve the distortions induced by industrialization policy, however, basic human needs policies compounded them still further. Basic human needs policies, after all, required expansion of the government consumption sector. In so doing, these policies reduced the share of private consumption in local economies below already involuntarily low levels; investment levels have typically remained inviolate. Thus, basic human needs programs have addressed the problem of local deprivation by the further separation of the individual in low-income society from the ability to identify and satisfy one's own needs directly.

The ostensible rationale for suppressing private consumption and augmenting gross domestic investment through development policy in low-income countries derives from its presumed prospects for accelerating economic growth. One may wonder why compulsory measures should be needed to bring about such a shift of resources; even low-income populations will voluntarily forgo consumption and increase their pool of investable savings if they judge the prospective rewards for such action to be appropriate. In theory a forced increase in investment levels might indeed open the possibility for quickening the pace of material progress. The record of the past generation, however, raises serious questions about the efficacy of this strategy in actual practice.

In some developing countries—Singapore, Korea, and Indonesia, among others—high rates of investment have indeed coincided with and have been followed by rapid rates of measured increase in national output. But in quite a few other countries, high rates of investment did not evince comparable results. In a number of countries, relatively high levels of investment have proven to be consistent with long-term economic stagnation—or worse.

A few examples may suffice. In the mid-1960s, according to the estimates of the World Bank,[6] the ratio of gross domestic investment in such countries as Nicaragua, Bolivia, and the Central African Republic was higher than in the United States. During the following two decades, however, per capita output in these countries, by World Bank estimates, registered a decline. That is to say: their per capita growth rate was negative. Zambia and Jamaica, according to World Bank estimates, also enjoyed substantially higher ratios of gross domestic investment to national output than the United States in the mid-1960s; they too registered negative per capita growth over the subsequent two decades. Rapid rates of population growth do not entirely explain these results, as the World Bank's estimates for aggregate growth of national output (GDP) for these countries demonstrate. As is well known, the American economy was characterized by comparatively slow growth between the mid-1960s and the mid-1980s. Yet by the World Bank's estimates, aggregate growth in the countries mentioned above was markedly slower than in the United States. By those numbers, Zambia's rate of growth in total output was less than half that of the United States; Jamaica's was scarcely a third of the U.S. rate.

Specific estimates of the International Monetary Fund differ from those of the World Bank; the picture they paint nevertheless is strikingly similar. According to IMF numbers, the gross domestic investment ratio for Africa as a whole has matched or exceeded that of the United States virtually every year since 1965. In view of the continent's by now familiar record of performance over the past decades, the comparison would seem to indicate strikingly low rates of return for a region wherein capital is typically thought to be scarce.

As we have noted, there are reasons to be careful in the use of such international economic statistics. These World Bank and IMF numbers, however, tell a story that would not be altered even by major statistical revisions or adjustments. The fact is that the investment sector in many low-income countries has generated shockingly low rates of return for many years. In such countries, the funds that are labeled productive capital have evidently been used in unproductive ways.

In its basic outline, the pattern of development manifest in many low-income countries over the past generation would seem to present the student of economics with a riddle. In many of these countries, rates of return on capital have evidently been low (in some cases possibly negative) for many years, even decades. In the face of competitive economic forces, one would have anticipated a decline in the investment ratio or an increase in rates of economic return—or both. How then are we to explain the instances—and they are numerous—where investment ratios appear not only to have maintained themselves but even to have increased in the face of persistently low rates of return?

International Capital Flows

One possible solution to this riddle would involve infusions of external capital. Under certain circumstances international capital transfers could help sustain the unstable equilibrium that seems to have characterized patterns of production in many low-income countries over the past generation. Admittedly the conditions required for solving this riddle through international capital flows not only would be highly restrictive but also might seem rather implausible. For they would necessitate not only steady and substantial inflows of capital to the countries in question, but also a formidable inattention, over years or even decades, to the uses to which international capital was being applied.

In the decades since President Truman announced America's Point Four program for economic aid to underdeveloped areas, Western countries and the international institutions they finance have indeed effected a steady transfer of resources to the developing world. The magnitude of this transfer is not commonly appreciated—even by students of economic development.

The Organization for Economic Cooperation and Development produces annual estimates of the net disbursement of financial resources to the low-income countries. These figures provide a basis for a rough estimate of the postwar transfer of capital from the West to the third world. Adjusting the OECD series for inflation, the net flow over the years 1956 to 1986 would amount to about $1.8 trillion in 1989 U.S. dollars. Note that these OECD estimates measure *net* disbursements—the residual after financial withdrawals, profit repatriation, and loan repayments have been taken into account (table 9–4).

How much is $1.8 trillion? A few comparisons may help to put the figure in perspective. It is almost three times the entire net worth of the U.S. farm system in 1987, according to estimates of the U.S. Department of Agriculture.[7] It is almost twice the value of all U.S. assets overseas in 1987, by the estimates of the U.S. Department of Commerce.[8] By the estimate of the U.S. Internal Revenue Service, it is

TABLE 9–4
NET WESTERN FINANCIAL FLOWS TO DEVELOPING COUNTRIES, 1956–1986
(constant U.S. billions of dollars, May 1989)

	1956–60	*1961–65*	*1966–70*	*1971–75*	*1976–80*	*1981–85*	*1986*	*Total*
Total net disbursement	126.5	143.9	195.8	298.3	511.1	442.4	73.8	1,790.8
Direct private investment	34.6	31.4	42.5	NA	49.6*	49.3	13.2	NA
Direct private investment (% of total net disbursement)	27.4	21.8	21.7	NA	11.9*	11.1	17.8	—

NA = not available.
* = 1977–80. Series adjusted by the U.S. Producer Price Deflator.
— = not applicable.
SOURCES: Organization for Economic Cooperation and Development, *Geographical Distribution of Financial Flows to Developing Countries* (Paris: OECD), various issues; OECD, *The Flow of Financial Resources to Less Developed Countries* (Paris: OECD), various issues; U.S. Bureau of the Census, *Statistical Abstract of the United States: 1988* (Washington, D.C.: U.S. Department of Commerce, 1981), p. 444.

more than twice the net worth of all U.S. millionaires in 1982—the most recent year for which such estimates have been prepared.[9] And as of the end of 1988, $1.8 trillion amounts to nearly three-quarters of the total market value of all the companies traded on the New York Stock Exchange.[10]

This is far from a precise estimate of the net transfer of wealth from the West to the developing countries. There are obvious technical problems in attempting to compute this flow of resources and theoretical ones as well. It is difficult to select an appropriate price deflator, or to adjust for fluctuations in the strength of the dollar. There is good reason, however, to suspect that the rough figure of $1.8 trillion may actually understate the net transfer of capital to the low-income countries during the postwar period. For one thing the OECD series does not attempt to include such items as private charitable transfers or military aid. And the figures in table 9–4 obviously do not include transfers in the early 1950s or in the years since 1986. If anything, the net transfer from the West to the developing world over the postwar period is likely to be even greater than the $1.8 trillion figure suggests.

The composition of the capital transfer from the West to the developing world over the past generation is quite interesting. The fraction of the flow deriving from direct private investment was never predominant, and over the 1960s, 1970s, and 1980s that fraction declined appreciably. The absolute volume of net direct private investment from abroad, after adjusting for inflation, was actually lower in 1985 than it had been in 1956.

In recent years the OECD has provided some estimates of the composition of net financial flows not only for developing countries as a whole but for categories of countries within the group. Some of these figures are presented in table 9–5. The breakdowns are intriguing. For the group classified as least-developed countries—a category with a population of approximately 1 billion persons—scarcely 5 percent of the total net flow of resources from the West was private in nature between 1977 and 1986. Even that figure exaggerates the private share insofar as it includes government-guaranteed export credits. Less than 1 percent of the total flow of Western funds to these countries over that same decade is estimated to have come from direct private investment. In 1986 private flows of capital to the countries in this group were recorded as negative. Concessionary public transfers accounted for more than 100 percent of the measured net resource transfer from abroad during that year.

The composition of resource flows to the group classified as middle-income countries—a category defined to include most of the members of the Organization of Petroleum Exporting Countries, as

TABLE 9–5
STRUCTURE OF NET WESTERN FINANCIAL FLOWS TO SELECTED CLASSIFICATIONS OF COUNTRIES, 1977–1986

	Least-developed Countries		*Middle-income Countries*	
	% of direct investment	% of private sector	% of direct investment	% of private sector
1977	2.7	10.0	16.2	58.3
1978	1.0	4.0	16.8	65.9
1979	0.5	8.9	18.0	62.1
1980	0.2	11.2	20.0	58.5
1981	1.5	5.1	19.1	60.7
1982	1.9	7.8	14.0	60.1
1983	0.5	1.9	11.3	51.5
1984	0.4	1.7	14.6	60.9
1985	0.4	1.4	20.2	0.1
1986	–0.3	–0.9	27.7	28.5
Unweighted averages				
1977–81	1.2	7.8	18.0	61.1
1982–86	0.6	2.4	17.6	40.2

NOTE: Least-developed countries are listed according to the UN classification system; country membership may change from year to year.
Middle-income countries are classified by World Bank criteria; group includes lower-middle-income and upper-middle-income countries for years 1979–1986, and middle-income, NICs, and OPEC-II groups for 1977–1978 in OECD typology; membership may change from year to year.
SOURCE: Organization for Economic Cooperation and Development, *Geographical Distribution of Financial Flows to Developing Countries* (Paris: OECD), various issues.

well as the so-called newly industrializing countries (NICs)—is substantially different. Even so, this composition may be somewhat surprising. Over the decade for which such data exist, concessionary public transfers accounted for almost half the measured inflow of capital from the West. Direct private investment accounted for less than a fifth of the net disbursement to this group over that decade. Direct private investment accounted for a smaller fraction of the net capital flow to middle-income countries in the early 1980s than it did to low-income countries in the late 1950s.

The relative eclipse of direct private investment as an instrument of capital transfer to developing countries is significant. Direct private investment is voluntary, and sensitive to prospects of economic opportunity. Over the past generation, low-income countries as a group have experienced considerable economic growth. It is anomalous that the relative (and by certain comparisons even the absolute) level of direct private investment should diminish, within the flow of resources to the developing world over these same decades. Despite economic advance the developing world as a whole appears to be less capable of attracting direct private investment than in the past—or more hostile toward it. This does not augur well for rates of economic return on the capital flowing into developing areas.

Comparatively little of the capital transferred to developing countries over the past generation has gone directly to private concerns or to individuals. Overwhelmingly, Western transfers flowed to the governments of low-income countries. By and large, the vehicles for these transfers have been outright grants, soft or subsidized loans, and commercial lending.

These vehicles of capital transfer have implications for economic performance. By definition, a soft loan carries an interest rate below the competitively prevailing rate of interest. A recipient can repay a soft loan even while achieving comparatively low rates of return. Often soft loans carry negative real rates of interest; in such circumstances a borrower can use funds in ways that actually diminish the capital and can still redeem contractual obligations. With an outright grant it is possible to deplete the gift completely without reducing the recipient's preexisting stock of capital.

Terms of public borrowing for low-income countries can help describe the economic environment in which local governments operate. In 1987 (the most recent year for which the World Bank's *World Development Report* provides figures),[11] the average interest rate for loans contracted to the government of Ghana was 1.9 percent; for the Central African Republic the rate was 1.2 percent; and for the government of Tanzania the rate was 1.2 percent. For Sierra Leone the interest rate for public loans contracted in 1987 on average was zero. These are nominal rates. After adjusting for inflation, real rates of interest on such loans were actually negative.

In effect, through such highly concessional transfers of capital, governments in low-income regions are being held to a lower standard of economic performance than those facing their own citizens, international businesses, or the governments of Western countries. Not all governments in developing regions have taken advantage of such relaxed standards of performance—but many have. If additional corroboration were needed, the so-called third world debt crisis of the 1980s eloquently supplies it. To the extent that it may accurately be described as a crisis, it is a crisis of productivity. Governments tend to find debts more difficult to repay when borrowed funds are generating extremely low rates of return.

The governments of many developing countries are currently experiencing difficulties with their external debt. Even more noteworthy, external government debt in a number of these countries consists principally of public rather than commercial loans. Public loans are typically concessional, and often they are highly so. To have difficulty meeting such terms reveals much, albeit inadvertently, about the policies and practices of the recipient government.

Conclusions

The preceding pages have not provided anything like a systematic analysis of the impact of ODA on recipient economies in low-income regions. Nor were they intended to do so. Instead they provide a brief review of what one might call the forensics of overseas development assistance.

Peculiar and pronounced distortions that are referred to elsewhere as "industrialization without prosperity" and "investment without growth"[12] now characterize the economies of many low-income countries. Prejudicial to the immediate well-being of local populations, these distortions also look to be financially costly in many instances. Because of the evidently low rates of return typically associated with them, the policies that created these distortions would appear to be inherently unsustainable in an open and competitive economic environment. Yet for decades they have been sustained. Massive transfers of capital from abroad, typically applied in ways that generated uncompetitive rates of return, could explain this continuing pattern of development. As we have seen, the transfer of resources from Western sources to low-income countries has been enormous over the past generation, and there is evidence that a significant portion of this transfer was used in decidedly unproductive ways.

The concessional public transfer of capital raises the possibility that recipient governments may adhere to separate, and lower, standards of performance than other economic actors. Obviously a government's performance and conduct need not be degraded by such bequests—but it can be. A government's behavior is most likely to be adversely affected by such bequests if that government wishes to be held to a less-exacting standard. After all, concessional transfers increase a government's freedom to maneuver—and thus to pursue its intentions. A government's intentions are thus central to the likely economic impact of overseas development assistance. If a government's intentions are economically untoward, overseas development assistance may actually finance programs so economically reckless, destructive, or injurious that they could not otherwise proceed. Unfortunately, the record of the past generation suggests this is more than a purely hypothetical possibility.

Development economics has devoted enormous attention to the theoretical possibilities of accelerating economic growth or reducing the incidence of local poverty through continuing bequests of ODA. It has devoted much less consideration to the actual record of use to which existing governments in low-income countries have directed such funds. Redressing this imbalance would seem to be not only in order but in the interest of the presumed beneficiaries of such study: the populations of the low-income countries themselves.

10

The Debt Bomb and the World's Children

In late 1989, for the second year in a row, the United Nations Children's Fund issued a dire and alarming assessment of the condition of, and prospects for, the world's children. According to these reports, a major—perhaps the major—hazard threatening children in less-developed countries (where the great majority of the world's children are to be found) arises from the forces of international finance. By this interpretation, what has come to be called the "third world debt crisis" directly imperils the well-being, and even the lives, of millions of children around the globe today.

The debts to which UNICEF refers have not been contracted by the children in question or by their parents. They are not owed by any private individuals. Instead they have been incurred by governments.

Over the past two decades sovereign states throughout the low-income regions have arranged to borrow great sums of money from abroad. By one authoritative estimate, the long-term external debt owed by governments and governmental organizations from developing countries approached $1 trillion (U.S.) at the end of the 1980s.[1] As the 1980s progressed, many of these governments failed to meet the scheduled terms of repayment. (The inability, or unwillingness, of a government to honor such contracts has come today to serve as the defining characteristic of a "debt crisis" and is also its proximate cause.) Now, UNICEF warns, millions of blameless children will suffer and die if international lenders should insist upon full payment of loans they extended to third world governments.

UNICEF's warnings have taken their most comprehensive form in *The State of the World's Children,* the survey and analysis it issues each year. In the 1989 edition UNICEF estimated that more than half a million children under the age of five were dying in the developing world every year as a direct result of the ongoing debt crisis.[2] According to the 1990 edition of *The State of the World's Children,* "poverty, malnutrition, and ill health are advancing again after many decades of retreat," and "the *sine qua non* of a resumption of human progress" in many regions of the world is now "some significant reduction in debt servicing."[3]

In UNICEF's analysis, "the poorest and most vulnerable children have paid the third world's debt with their *health* . . . the poorest and most vulnerable children have also paid the third world's debt with the loss of their opportunity to be *educated*" [UNICEF's italics].[4] Deteriorating health and educational conditions, in this reading, are largely caused by cutbacks in services and other adjustments in government policy designed to maintain international creditworthiness and to facilitate payments on overseas debts. In many countries, UNICEF concludes, "millions of children" are today "unnecessarily exposed" to "the sharpest edges of the adjustment."[5]

UNICEF may be the most authoritative and articulate voice today suggesting that debt-servicing obligations of governments in low-income areas are prejudicial to the interests of the poor and the vulnerable, but it is hardly alone. That view is widely shared by contemporary students of what is called development policy. It is also evident in the statements of many world leaders. To cite just one example, President Robert Mugabe of Zimbabwe was recently quoted as saying that "few scourges in human history can claim so many victims as today's debt crisis."[6] Nor are such statements peculiar to the leaders of low-income debtor countries. Almost identical quotes could be adduced for leading figures from the industrialized, aid-giving world.

If this interpretation is correct, its implications would be far-reaching and consequential. In recent years scheduled obligations by governments of developing countries have indicated repayments to international creditors averaging around $150 billion a year (including both interest and principal). If such payments directly reduce the well-being of children and other vulnerable groups in debtor countries, a general and immense affliction would seem to have been routinized into the ordinary workings of the current international order. Redress of the injuries wrought by these arrangements would presumably require vigorous remedial measures and might ultimately entail a bill of hundreds of billions of dollars. No less distressing would be the corollary to such requisite therapies, namely, the conditions of children and other vulnerable groups had been permitted to deteriorate even though representatives of both low-income governments and international organizations have been authorized, and instructed, to speak on their behalf and to guard their interests.

It is impossible to do justice to the entire range of analysis and arguments associated with UNICEF's current viewpoint in a single, brief statement. Rather than attempt to do so, I instead examine three issues that are central to the interpretation of events suggested in the most recent editions of *The State of the World's Children.*

The first is the proposition that existing international financial arrangements are currently draining the developing world of tens of billions, or even hundreds of billions, of dollars each year, resulting in a net transfer of wealth from poor countries to rich countries. The second is the proposition that health and educational conditions for children in low-income regions have demonstrably deteriorated in recent years, after decades of general progress. The third is the proposition that government policy adjustments in low-income countries, in response to international financial pressures, have evolved in such a manner as to diminish systematically the capability of societies and states to attend to the needs and problems of vulnerable and distressed groups. Let us examine the evidence concerning each of these propositions in turn.

Resources from Poor to Rich?

Among students of development and financial specialists alike, today's prevailing perception is that net resource transfers are flowing from poor countries to rich ones. In this view, the combination of several factors—floating international interest rates, reduced willingness to lend to low-income governments, and diminishing commitments of official development assistance (ODA, or foreign aid)—resulted in a signal shift in the direction of resource transfer during the 1980s. By this argument, whereas postwar transfers to the developing world had been strongly positive, at some point—perhaps as early as 1982—the direction of flows was reversed. Between 1980 and 1988, according to a recent World Bank estimate, long-term debt-service payments for developing countries nearly doubled, rising from about $76 billion to about $133 billion;[7] repayment requirements for shorter-term debts may have risen even more dramatically. New inflows from foreign aid, private investment, and overseas loans were inadequate, by this reading, to cover such obligations. In 1987 alone, according to the World Bank's 1989 *World Development Report*, the net transfer abroad of financial resources by developing countries approached $40 billion.[8]

With the hardening of loan conditions toward developing countries, it is argued, the magnitude of that ostensibly regressive transfer may have increased subsequently.

The notion that the third world is being drained of resources by rich countries is not a new one, although it appears to enjoy much more currency today than in the recent past. As with other opinions concerning empirical matters, however, the facts rather than the opinions are ultimately immutable. Close examination of some of the particulars of the international financial picture in the 1980s suggests that the actual circumstances facing less-developed countries were considerably different from those commonly depicted.

Measurement of international financial flows and transfers is, necessarily, an inexact business. Ascertaining the precise volume of flows by category from one country to another is never easy; even under the best of circumstances such estimates are likely to be incomplete, and thus not fully accurate. Standardizing transactions by converting them into a single currency, whether for a given year or over time, is a problematic task, even in theory; there is no single correct way to do so. Categories often seem arbitrary (think only of "capital flight," today's term for voluntary overseas investments by private individuals from low-income countries whose business climates are deemed unfavorable.) The method one adopts in valuing international flows, moreover, can directly affect results: different implications might be suggested by an accountancy approach, which would compute the flow of funds on a direct accrual basis, and by an economic approach, which would discount or adjust such flows to their net value in accordance with such things as the competitive cost of capital. Given these and other complexities, discrepancies between estimates, and even errors within estimates, can be almost taken for granted.

Some errors, however, are avoidable. In 1988 the Organization for Economic Cooperation and Development conducted a comprehensive review and analysis of available data on financial transfers to developing countries. OECD was well positioned for such a study, insofar as it has been compiling such data since the 1950s and has for several decades published the authoritative annual report, *Geographic Distribution of Financial Flows to Developing Countries*. Its conclusions are worth quoting at length:

> Much of the public concern . . . about the turn-around in the development financing picture between the 1970s and the 1980s has been based on various estimates for net financial transfers which suggest that finance has been flowing from poor countries to rich countries rather than vice versa. The net financial transfer concept is an arithmetical construct which

> needs very careful interpretation to yield meaningful conclusions. . . . a *positive* net financial transfer (net capital receipts exceeding interest payments) provides a cash flow which enables imports to exceed exports and thus expands domestic investment and consumption possibilities beyond the level of national output. . . . Interpreting the economic concept in this latter way makes it clear that grants which expand domestic consumption and investment possibilities (such as ODA grants) and direct investment should be counted equally with the flows in the net financial transfer equation. *Estimates of net financial transfers to developing countries which are restricted to debt financing (. . . for example the commentary accompanying the World Bank's debt tables and* World Development Report) *give an overly pessimistic view of the current situation* [emphasis added].[9]

In other words, some of the most commonly cited estimates on financial transfers to and from less-developed countries had systematically neglected to include major sources of overseas finance in their calculations—sources, moreover, in which the transfer to developing regions was unmistakably and significantly positive.

The World Bank has just recently reacted to such comments and criticisms by revising its estimates. In December 1989, only a few months after the *World Development Report* mentioned above had appeared, the bank published a new and dramatically different set of figures on net financial transfers to developing countries during the 1980s. For 1987, the bank still estimated that the developing countries had experienced a net drain of financial resources, but whereas the previous number had been almost $40 billion, it was now under $10 billion. For 1984, the bank had previously calculated that the developing countries had lost about $10 billion in the process of net financial transfers; the new computations suggested that the region as a whole had actually enjoyed a net *inflow* of almost $20 billion.[10] Under scrutiny the World Bank evidently changed not only the magnitude of its estimates on these flows but also their direction.

While these recent revisions eliminate some of the more obvious problems with the World Bank's third world debt estimates, additional difficulties may remain. It is useful for that reason to refer to the perhaps more careful and comprehensive estimates prepared by OECD (table 10–1). According to these figures the net financial transfer to developing countries, whether measured in current or inflation-adjusted terms, did indeed decline substantially after 1982. Even so, the net transfer to the third world by these numbers appears to have remained strongly positive. From 1983 through 1987 (the most recent year for which figures were available) the estimated annual net financial transfer

TABLE 10–1
ESTIMATED NET FINANCIAL TRANSFERS TO LESS-DEVELOPED COUNTRIES, 1980–1987
(current U.S. $ billions)

Region	*1980–82 Average*	*1983*	*1984*	*1985*	*1986*	*1987*[a]
Sub-Saharan Africa	12	10	8	10	13	16
North Africa and Middle East	9	5	7	8	6	4
Asian low-income countries	12	12	14	17	17	22
Other Asia	6	9	1	–3	–2	–2
Western Hemisphere	30	–8	–8	–15	–10	–4
Other transfers and adjustments[b]	8	–2	5	3	3	–2
Total LDCs[c]	77	27	27	20	27	34
Number of least-developed countries	10	9	9	10	12	13

NOTES: Regions as defined by OECD.
a. Provisional.
b. Europe, Oceania, unallocated, and other adjustments.
c. Excluding gulf countries and Taiwan.
SOURCE: Organization for Economic Cooperation and Development, *Development Cooperation: 1988 Report* (Paris: OECD, 1988), p. 53.

to developing countries in current terms ranged from $20 billion to $34 billion. For these five years alone, in this reckoning, the cumulative net financial transfer to the less-developed countries, far from being negative, was actually a positive balance of $135 billion.

According to these figures, moreover, the two poorest areas of the third world—sub-Saharan Africa and low-income Asia—enjoyed not only strongly positive but increasingly positive net transfers since the onset of the third world debt crisis. Correspondingly, net transfers to countries designated as least developed are calculated to have been greater in the late 1980s than in the years before the debt crisis.

By these estimates, the western hemisphere—Latin America and the Caribbean—did indeed experience net negative transfers between 1983 and 1987. Interestingly enough, however, the category of countries described as other Asia also shows a negative net transfer from 1985 onward. The negative figures for that region, however, do not indicate financial distress. Quite the opposite: countries like South Korea have been achieving substantial balance of payments surpluses and are now retiring their international debt. Others in the area, such as Taiwan, are now actually net international creditors.[11] If a country is ever to make good on its international debts, it must at some point transfer more capital overseas than it is bringing in from abroad. To put it another way, it must "suffer a net negative financial transfer" someday if its debts are ever to be redeemed.

Both Korea and Argentina apparently experienced negative net financial transfers in the late 1980s. Korea's policies and practices, however, seem to have resulted in the highly productive use of borrowed funds and left its government creditworthy. Argentina's did not. The occurrence of negative net transfers is therefore less significant than the reasons for them. By the same token, positive net transfers need not indicate the economic health of a country or the financial competence of its government. As OECD cautiously has put it, "the case of Korea . . . illustrates the need for careful interpretation of the net financial transfer concept."[12]

So much for the arguments about financial flows in the less-developed countries. What about the widespread perception that terms of lending to the poor countries have hardened severely since the advent of the debt crisis?

Here again data and computations from OECD are instructive. OECD estimates the annual "interest cost" on long-term debt for developing countries; it also breaks these estimates into three broad groupings corresponding to the country's estimated income level. (The concept of interest cost is not precisely the same as that of interest rate, but for our purposes calculations produced by the two are close.) The interest costs for long-term third world debt may be compared with the London interbank offer rate (LIBOR) on long-term loans in dollars.[13] This rate applies in interbank transactions for banks of the highest creditworthiness, and may be taken as an approximation of the interest rate for customers who are deemed to be virtually zero-risk borrowers. Comparison of these two sets of numbers should help to indicate the extent of concessionality, if any, in lending to low-income governments in recent years (table 10–2).

According to these estimates and data, total interest costs on long-term debt for developing countries as a whole have never been higher than the corresponding LIBOR rate for long-term dollar loans and usually have been considerably lower. For those groups of countries designated as low income and lower-middle income—the poorest country groupings—the LIBOR rate is greater than the interest cost for every year from 1980 through 1988 (the latest year for which such

TABLE 10–2
CONCESSIONALITY IN LONG-TERM LENDING TO LESS-DEVELOPED COUNTRIES BY INTEREST COSTS VERSUS LIBOR-DOLLAR ONE-YEAR RATES, 1980–1988
(percent)

Category	*1980*	*1981*	*1982*	*1983*	*1984*	*1985*	*1986*	*1987*	*1988*
Total LDC long-term debt	8.9	9.5	9.6	8.0	7.8	9.0	7.0	5.7	6.0
Low-income countries	4.4	5.0	5.2	5.2	4.5	5.4	4.6	4.3	4.3
Lower-middle-income countries	8.4	8.0	8.0	7.2	7.2	7.6	6.3	5.7	5.6
Upper-middle-income countries	10.8	11.5	11.5	8.9	9.5	10.3	8.0	6.5	7.2
LIBOR-dollar one-year rate	13.4	16.1	13.7	10.2	11.8	9.1	7.0	7.6	8.4

NOTE: Interest costs for less-developed countries are defined as annual interest effective payments and other charges (including spreads and fees on floating interest debt) as a percentage of disbursed debt at the beginning of the year. Long-term debt is defined as obligations with a maturity of more than a year. Subcategories for less developed countries as classified by the United Nations and the World Bank; because of changes in income groups, the data for 1980–1983 are estimates of OECD. LIBOR refers to London Interbank Offer Rates.
SOURCES: For LDC interest costs: Organization for Economic Cooperation and Development, *Financing and External Debt of Developing Countries: 1988, Summary* (Paris: OECD, 1989), p. 64. For LIBOR rates: International Monetary Fund, *International Financial Statistics* (Washington, D.C.: May 1990), p. 66, and 1983 *International Financial Statistics Yearbook* (Washington, D.C.: IMF, 1983), p. 66.

figures are available). In and of itself, this spread indicates concessionality. Lending to countries in these groupings appears to have been substantially subsidized—and to continue to be subsidized. A significant component of the "loans" these country groups contracted during the 1980s were in fact gifts or grants. Concessionality in lending to these governments, one may note, continued to be characteristic after the advent of the third world debt crisis.

For the group of countries designed as upper-middle income, the picture appears to be somewhat more complex. From 1980 through 1984, their average interest costs were lower than the long-term LIBOR dollar rates. Lending during those years, then, was on the whole concessional. In the two most recent years for which data are available, 1987 and 1988, lending to this group was also conducted on an ostensibly concessional basis. In 1985 and 1986, however, interest costs for this group were higher than the corresponding LIBOR rates.

One may argue that lending was not characterized by concessionality during those two years. On the other hand, 1985–1986 was a period of major defaults and "reschedulings" by a number of governments within this country grouping; such activity did not improve credit ratings. Thus one might equally conclude that lending to upper-middle-income countries continued to be concessional during those years—insofar as the actual risk premium for lending to these countries was arguably far larger than the spread between interest costs and LIBOR rates would have suggested.

The degree of concessionality in long-term lending to developing countries does appear to have been lower at the end of the 1980s than at the beginning. In 1988, the average interest costs for long-term loans to governments in developing countries were about two and a half points lower than the best possible rate for a preferred bank in the West; in 1980, by contrast, that spread was about four and a half points, and in 1981 it was more than six and a half points. The early 1980s, however, was an unusual and atypical period in international finance. With oil shocks, rapid rates of international inflation, and negative real interest rates at the central banks of the major Western economies, those years were in no sense representative of the previous, or subsequent, experience in the international economy.

In the early 1970s, for example, the element of concessionality in lending to developing countries was much less pronounced than a decade later. Lending to developing countries was also apparently more concessional in the late 1980s than it had been in the early 1970s.[14] Only against expectations framed during the late 1970s and early 1980s would current levels of concessionality in long-term lending to less-developed countries seem ungenerous.[15]

So-called debt crisis notwithstanding, the net flow of funds to developing countries as a whole appears to be strongly positive, and lending appears to remain highly concessional. To the extent that there may be said to be a general crisis in state finance in less-developed countries today, it may be more closely related to governments' expectations, attendant policies, and corresponding economic results than to any changes in the actual availability of financial resources from abroad.

Setbacks in Health and Education?

It is widely believed that living conditions for vulnerable groups, including children, have severely deteriorated in many regions of the low-income world during the 1980s. That deterioration, some believe, has included broad and serious setbacks in health and educational conditions. UNICEF clearly subscribes to this belief. According to the 1990 edition of *The State of the World's Children,* "infant mortality is known to have risen in parts of Latin America and Africa south of the Sahara."[16] Moreover, according to UNICEF,

> in the mid-1980s the proportion of children enrolled (in school) began to fall while the total numbers of children continued to grow. As a result, the number of children [aged six–eleven] out of school has increased [in 1987] to 60 million—the first significant rise in four decades.[17]

The implications of such dramatic statements are obvious. What is less obvious, however, is how such dramatic statements are substantiated.

Take the question of mortality trends. To substantiate a rise in infant mortality rates for any country over the past several years, one would need to have a reasonably complete registration of births and infant deaths, preferably released on at least an annual basis.[18] Exceedingly few low-income countries, however, have reasonably complete registration data for births and deaths.

The United Nations Population Division (a unit within its Department of International Economic and Social Affairs) has been compiling and analyzing international demographic data since the 1940s and has published the authoritative *Demographic Yearbook* since 1948. According to the 1988 *Demographic Yearbook* only fourteen countries or territories of a million persons or more in the third world have "virtually complete" infant mortality registration (90 percent completeness or better). Those fourteen places have an estimated population of about 105 million people as of 1988. They would, by the UN's reckoning, thereby account for less than 3 percent of the total population for the developing countries for that year. None of these places is in sub-Saharan Africa. Eight are in Latin America or the Caribbean—but one of these is Cuba, which presumably does not suffer from the problems of the capitalist world, and five of the others do not appear to be publishing infant mortality data on a regular annual basis. Thus, for countries composing well more than 95 percent of the estimated population of the third world, it is not immediately apparent how one would go about documenting recent year-to-year changes in infant mortality rates.

The UN's Population Division does produce periodic projections and estimates of infant mortality rates for the various countries and regions of the world—with the caveats and disclaimers appropriate for such an exercise. In 1983 it issued global projections of infant

TABLE 10–3
ESTIMATED INFANT MORTALITY RATES, 1980–1985
(per 1,000 live births)

Region	*Projected in 1982*	*Estimated in 1986*	*Difference (%)*
World total	81	78	–4
More-developed regions	17	16	–6
Less-developed regions	92	88	–4
Africa	116	112	–3
Eastern Africa	109	120	+10
Middle Africa	118	117	–1
Northern Africa	104	100	–4
Southern Africa	94	87	–7
Western Africa	135	123	–9
Latin America	63	62	–2
Caribbean	58	65	+12
Central America	56	57	+2
Temperate South America	42	32	–24
Tropical South America	70	69	–1
East Asia	38	36	–5
South Asia	109	103	–6

NOTE: Regions defined according to United Nations classification.
SOURCE: United Nations, *Mortality of Children under Age Five: World Estimates and Projections, 1950–2025* (New York: UN Department of International Economic and Social Affairs, 1988), p. 25.

mortality rates for the period 1980–1985. In 1988 it published revised estimates based on a review and analysis of subsequently available data. The projections and estimates are compared in table 10–3. On the whole, infant mortality rates for the developing world are lower in the revised series than in the initial projections. (The only significant exceptions to this generalization are Eastern Africa, which was affected by largely manmade famines, and the Caribbean, where new data suggested that infant mortality rates in Haiti and the Dominican Republic had been underestimated).[19] If the so-called debt crisis, whose onset is generally dated from Mexico's suspension of payments in 1982, had significantly affected infant mortality rates in developing countries, one might well have expected revised estimates of infant mortality to be generally higher than initial projections. In this exer-

TABLE 10–4
REPORTED RATES OF INFANT MORTALITY FOR NON-COMMUNIST LESS-DEVELOPED COUNTRIES, 1980–1988
(per 1,000 live births)

Country and Region	*1980*	*1981*	*1982*	*1983*	*1984*	*1985*	*1986*	*1987*	*1988*
Africa									
Mauritius	33.0	34.1	29.4	25.8	23.1	24.2	26.4	23.7	NA
Asia									
Hong Kong	11.2	9.7	9.9	9.9	8.8	7.5	7.7	7.4	NA
Kuwait	27.7	24.1	22.8	19.0	18.5	18.4	15.6	NA	NA
Peninsular Malaysia	24.9	21.0	20.4	20.3	17.5	17.0	15.5	14.4	NA
Singapore	11.7	10.7	10.7	9.4	8.8	9.3	9.4	7.4	7.0*
South America									
Chile	33.0	27.0	23.6	21.9	19.6	19.5	19.1	18.5	NA
Uruguay	37.0	33.4	29.4	28.3	30.1	29.4	28.0	23.8	NA

NOTES: Table 10–4 includes all non-Communist less-developed countries, according to estimates by sourcebook, with populations at 1 million or more for midyear 1988, with "virtually complete" infant mortality registration for 1980 through 1988 and with unrevised and unbroken series for the entire period. "Virtually complete" infant mortality registration is defined to mean registration of 90 percent or more of infant deaths. Regions as classified by *UN Demographic Yearbook*.
NA = not available.
* = provisional.
SOURCE: United Nations, *Demographic Yearbook* (New York: UN), various issues.

cise—perhaps the most careful and comprehensive of its kind to date—that did not occur.

These UN estimates are only that—estimates. We can also examine the infant mortality rates actually reported by less-developed countries with reasonably complete registration figures, with continuous and unbroken data series, and with estimated populations of 1 million persons or more as of midyear 1988. Only nine places in the low-income world are judged to conform to these specifications; one is Cuba. The remaining eight account for barely 1 percent of the population estimated for the developing world (table 10–4). One might reasonably argue that these places are unrepresentative of third world infant mortality trends precisely because they are capable of collecting such data on a reasonably reliable and annual basis. Whatever one may make of it, table 10–4 indicates a pattern of steady decline in infant mortality rates in these places over the 1980s.

In some countries reported rates of infant mortality do rise from one year to the next—as indeed they have from time to time in some of the countries of the postwar West. But just as in developed Western countries, such upticks at least to date look to be variations on an unmistakable downward trend. One may note that two of the countries in table 10–4—Chile and Uruguay—rescheduled their international debts during the 1980s. By doing so, they would currently be defined as debt crisis countries. In their particular cases, a debt crisis evidently did not forestall rapid reductions in infant mortality. Between 1980 and 1987, the reported rates of infant mortality dropped by more than a third in Uruguay, and by almost half in Chile.

Given the limitations of current information, one cannot state conclusively that infant mortality rates in recent years have continued their decline in each and every poor country, or in all of those whose governments have failed to repay their international debts on schedule. The general lack of data on infant mortality for poor countries, furthermore, may suggest that governments in low-income regions are generally either not capable of, or interested in, charting the condition of their most-vulnerable groups with any great accuracy. In and of itself, this would not seem to augur well.

Nevertheless, at least to date, there appears to be no serious evidence to suggest that the broad decline that has characterized third world infant mortality trends since the end of World War II has been arrested, much less reversed. The assertions that infant mortality rates have been on the rise in African and Latin American countries since the beginning of the debt crisis is, quite simply, something other than an empirical conclusion.

What of the contention that the proportion of third world children enrolled in schools has been declining since the onset of the debt crisis? Here figures prepared by the UN Educational, Scientific, and Cultural Organization may be helpful. UNESCO estimates both gross enrollment—the total number of students in relation to the population of school age—and net enrollment—actual proportions of the age groups in question actually in school. The distinction is crucial in many developing countries, where grade school classes often include teenagers and sometimes even adults.

One should be aware of the margins of error attendant upon these estimates. For many less-developed countries, there are uncertainties not only about the numbers of children actually enrolled in schools but also about the size of the cohorts from which they are drawn. The decimal points in UNESCO's estimates are obviously unwarranted. False precision notwithstanding, these estimates are nonetheless instructive (table 10–5).

UNESCO separates children into two groups: ages six to eleven and twelve to seventeen, corresponding roughly to the periods for primary and secondary schooling. According to UNESCO calculations, slightly less than half of all children between the ages of six and eleven in developing countries were enrolled in school in 1960. By 1987 (the latest year in its most recent series) the proportion was estimated to have risen to a bit over three-fourths. This would be far from universal schooling, even at the lowest grades. Even so, it would be a higher rate of net enrollment for developing countries than ever before. The gap between net enrollment ratios in developed and developing countries, at least by UNESCO's reckoning, has never before been so small.

In Latin America and the Caribbean, Asia, and the Arab states, to use UNESCO's categories, net enrollment in the primary school ages was estimated to have risen steadily throughout the 1980s. For sub-Saharan Africa enrollment was estimated to have dropped by several points for boys and girls alike between 1980 and 1985, resuming a rise thereafter. Those numbers may speak to real reversals in educational opportunities for the region during that interval.

Government statistics, however, are more limited and less reliable in Africa than in any other region of the low-income world. It is not immediately obvious, for example, why enrollment ratios for sub-Saharan teenagers should be expected to rise if they were dropping for children of grade school age, yet this is precisely what UNESCO estimates suggest for the period in question. For children roughly of high school age, incidentally, UNESCO estimates indicate a steady rise in enrollment ratios in every region of the third world throughout the 1980s.

UNICEF's *State of the World's Children 1990* cites UNESCO for its claim that enrollment ratios have been generally dropping in poor countries. Yet UNESCO's own figures suggest precisely the opposite. UNICEF's claim of an abrupt rise in the late 1980s in the numbers of third world children not in school looks equally curious. UNICEF does not explain exactly how it arrived at this calculation, and it is by no means obvious how one would replicate the procedures by which it was produced.

The absolute number of children not enrolled in school is a function of net enrollment and the number of school-age children. For UNICEF's posited rise to be valid, and consistent with the UNESCO figures, the number of children aged six to eleven in the third world would have needed to rise more than 25 percent between 1986 and 1987. For all

TABLE 10–5
ESTIMATED NET ENROLLMENT RATIOS BY AGE GROUP FOR CHILDREN IN DEVELOPING COUNTRIES, 1980–1987
(percent)

	6–11 years		*12–17 years*	
Country and Year	Total	Female	Total	Female
Developing countries				
1980	69.6	62.5	43.0	36.8
1985	74.2	67.4	43.8	37.7
1987	76.2	70.0	44.8	38.4
Africa (excluding Arab states)				
1980	58.5	52.5	42.9	34.9
1985	54.6	50.2	44.0	35.5
1987	55.8	51.1	45.8	37.3
Asia (excluding Arab states)				
1980	70.5	62.8	41.8	35.5
1985	77.2	69.6	42.1	36.0
1987	79.7	73.0	42.7	36.2
Arab states				
1980	68.1	57.9	42.9	33.0
1985	71.5	62.8	49.1	39.7
1987	73.4	65.0	50.8	42.0
Latin America and the Caribbean				
1980	82.4	81.9	62.6	61.6
1985	85.2	84.7	66.2	65.1
1987	86.3	85.7	68.2	67.2

NOTE: Decimal points as printed in source. Net enrollment refers to numbers of children of designated age groups actually enrolled in schools. Country categories as classified by UNESCO. Decimal points are found in the original estimates.
SOURCE: United Nations Educational, Scientific, and Cultural Organization, *Statistical Yearbook 1989* (Paris: UNESCO, 1989), pp. 2–34.

the talk of a population explosion in developing countries, such a jump is obviously a demographic impossibility.

As it happens, UNESCO's own numbers suggest that the total number of grade school–age children not enrolled as students in the third world would have fallen continuously during the 1980s as long as the cohort of children six–eleven years of age grew less than 3.5

percent a year. According to projections by the UN's Population Division—another source cited in the UNICEF calculations—that cohort was expected to grow well under 1 percent a year in less-developed countries, reflecting the significant decline in fertility in much of the low-income world in the 1970s and early 1980s. Whatever their accuracy, figures published by UNESCO and the UN's Population Division point not to a sudden sharp rise in the number of children in poor countries with no access to the grade school classroom, but rather to a steady and continuous diminution in the size of this disadvantaged group.

Even for those enrolled in schools, the quality of education is often decidedly poor. State policies toward education (and health) in low-income countries commonly leave much to be desired. The tendency to subsidize the privileged and to neglect the needy is prevalent. In Nepal, where barely a third of the nation's girls were thought to have been enrolled at appropriate levels in grade schools in 1984, the government was apparently devoting almost equal amounts to college and primary education. In the People's Republic of the Congo, the local government was apparently spending more on higher education than on primary schooling during the 1980s, even though conditions in the country's primary school system were such that UNESCO could not provide as much as a guess for enrollment levels.[20]

It is one thing to judge progress to be inadequate and quite another to deny its existence. Systematic misreadings of the condition of the world's children are unlikely to stand the young and the vulnerable in good stead. They may confuse or mislead groups actively concerned with the plight of the most distressed. They may convince citizens and public officials that manageable problems are unmanageably large. They may even discourage the adoption of measures and policies whose ultimate and predictable beneficiaries would include large numbers of children in the low-income world.

Adjustment Policies

By UNICEF's diagnosis, the "disastrous trends of the 1980s"[21] are directly related to the broad embrace by governments in low-income regions of what are today called adjustment policies. The term refers to a package of measures adopted to improve a government's budget discipline, financial stability, and creditworthiness. Typically adjustment policies are expected, and intended, to enforce a degree of austerity upon government expenditures. Many of the highly indebted and financially distressed governments in low-income regions in the 1980s were encouraged to embark upon adjustment by their creditors.

The State of the World's Children 1990 is sharply critical of most adjustment policies. "The harshness of the adjustment process, its political dangers, and its environmental consequences," the report states, "have not entirely escaped . . . notice."[22] "The problem," it explains, "is that education, health, birth planning services, and natural resource protection all involve government spending, a habit most government adjustment programs are seeking to discourage."[23] "The marketplace can be a brutal place for those who lack the purchasing power to make it serve their needs," it warns—"as the problems facing the children of today's free-market economies clearly show."[24] In UNICEF's view even concessional lending authorities may play a role in promoting such brutality for "at the moment, World Bank and IMF loans are conditional on the fulfillment of economic criteria such as reducing public spending and stabilizing the balance of payments."[25] In UNICEF's view, by contrast, "it is now essential to take a wider view of these conditions. . . . economic adjustment programmes need to take into account . . . *social impact*. . . . The days are gone when financial institutions could confine their concerns to the purely economic" [UNICEF's italics].[26] UNICEF recommends instead what it calls "adjustment with a human face." While all the practical details of this strategy have not been spelled out, the longest programmatic description of the idea to date commended the approach of the government of Peru—which had then summarily suspended most of its international debt payments in the name of economic growth and social welfare at home.[27]

By the World Bank's count, at least sixty non-Communist governments in low-income areas rescheduled some of their international debts in the 1980s.[28] Many of these governments announced austerity measures attendant upon the renegotiation of their debts. Moreover, numerous governments whose finances were not in disarray implemented deflationary adjustments and other stabilization policies during those same years. Did adjustment policies generally undermine the efforts, or capabilities, of governments in low-income countries to protect the poor and vulnerable in the wake of the debt crisis?

Data compiled by the International Monetary Fund can help us address this question. More than 100 governments in low-income Africa, Asia, and Latin America are members of the IMF and thereby required to share financial and economic information with the organization. On the basis of these data, the IMF estimates the ratio of government spending to national output around the world and over time. From this series table 10–6 shows the proportion of gross national product comprised of central government expenditures and new lending minus repayments. This measure in effect adjusts for the impact of the debt crisis, insofar as net payments by government abroad would reduce the estimated size of the government sector.

For a variety of reasons, these estimates are far from perfect. They follow only expenditure by the central government, even though local government and state-owned enterprises often figure prominently in state finance in low-income regions. Data on expenditures are often inadequate: less than a third of the governments of sub-Saharan Africa seem to provide the IMF with uninterrupted and up-to-date information on budgetary outlays. (The accuracy of such data is another matter altogether.) And figures on gross domestic product, as is well known, are frequently problematic. For all these limitations, the IMF's estimates are nonetheless revealing.

In 1980, according to IMF estimates, central government expenditures adjusted for lending and repayments amounted to slightly more than a quarter of output for developing countries as a whole. In the United States, by contrast, such spending accounted for less than a quarter of GDP in 1980. By the end of the 1980s, according to the IMF, that ratio was lower in the United States than it had been in 1980, or in any interim year. For developing countries as a whole, conversely, it appeared to be higher toward the end of the 1980s than at the beginning—higher, for that matter, than for any year in between. Even after adjusting for debt repayments, the ratio of government spending to GDP appears to be up sharply not only for such financially pressed groups as the oil-importing developing countries but for such regions as sub-Saharan Africa and the Western Hemisphere (that is, Latin America and the Caribbean). While year-to-year dips can be located in the series, the only broad region in which the ratio dropped substantially appears to be the Middle East, where expenditures have been constrained by the drop in oil revenues.

Even among the states most prominently featured in the debt crisis, no clear downward trend in the net share of central government spending is evident to date. In some places, such as Zambia, there looks to have been a substantial drop. In others, such as the Philippines and Argentina, figures are a bit higher, or lower, at the end of the 1980s than at the beginning. Still elsewhere, as in Brazil, the share of government seems to have risen rapidly and inexorably.

Generally speaking, the rise in government's share of the economy appears to have been more rapid between 1980 and 1982 than in subsequent years. By this timing, it would appear as if the debt crisis did affect government spending in much of the third world. That impact, however, seems to be widely misperceived. Far from suffering a general reduction in the size of government, low-income regions on the whole apparently experienced merely a slowdown from an earlier, extremely rapid tempo of government growth. Only in relation to the pace of expansion in the years leading up to the debt crisis would it

TABLE 10–6

ESTIMATED CENTRAL GOVERNMENT EXPENDITURES PLUS LENDING, MINUS REPAYMENTS, AS A PERCENTAGE OF GDP FOR SELECTED REGIONS AND COUNTRIES, 1980–1988

	1980	*1982*	*1984*	*1986*	*1988 (or latest year)*
Developing countries	25.6	27.8	25.7	27.9	28.0[a]
Africa	22.9	24.6	24.6	26.6	NA
Asia	20.7	21.4	20.5	23.3	22.4[a]
Middle East	37.6	40.1	35.0	34.0	NA
Western Hemisphere	22.0	27.4	24.5	28.5	30.1[a]
Non-oil developing countries	25.3	27.8	26.4	28.6	28.7[a]
United States	23.8	25.4	24.6	24.9	23.3[b]
Major debt-rescheduling states					
Ivory Coast	34.9	NA	31.7	NA	NA
Nigeria	NA	NA	14.6	19.2	24.5[a]
Zambia	44.3	42.8	31.0	NA	37.5
Philippines	14.3	15.4	12.4	17.5	16.5
Argentina	24.7	23.0	19.9	23.1	22.2[a]
Brazil	23.6	27.7	28.1	35.5	45.4[a]
Chile	28.7	31.2	31.9	29.3	29.7
Mexico	18.0	30.4	23.3	29.0	26.8
Venezuela	26.2	33.2	22.2	24.0	NA

NOTES: Regions classified by IMF.
NA = not available.
a. 1987.
b. 1989.
SOURCE: International Monetary Fund, *Government Finance Statistics Yearbook 1989*, vol. 13 (Washington, D.C.: IMF, 1989), pp. 94–95.

seem possible to argue that current third world state spending practices are characterized by austerity.

UNICEF argues that financial pressures have caused cutbacks in health and education programs throughout low-income areas. But programs for health and education have seldom enjoyed a high priority in third world budgets. Taken together, health and education are currently estimated to account for barely one-seventh of current central government expenditures for the developing countries (table 10–7).

In 1980, according to the IMF, central government commitments for health and education amounted to a little more than 3 percent of GDP for its low-income member governments. Between 1980 and 1987 the percentage of central government spending in GDP, net of lending and repayments, rose almost three points for developing countries as a whole and more than three points for the oil-importing countries. Thus, even after the onset of the debt crisis, it would have been possible virtually to double the share of public resources allocated to health and education in the third world without reducing the share for other activities—if presiding governments had wished to do so.

With few exceptions, third world governments did not choose to do so. In a number of countries, the actual pattern of allocations remains unclear. The IMF, for example, is apparently unable to provide estimates of public health and education expenditures for sub-Saharan Africa for any year of the 1980s and has figures on health and education for 1988 for only four countries in the region—facts that, perhaps, speak for themselves. In one or two countries public spending on health and education appears to have virtually collapsed: in 1987, for example, less than 4 percent of Nigeria's central budget was apparently earmarked for such purposes. On the whole, however, the share of government spending for health and education in poor countries seems to have remained fairly constant throughout the 1980s: rising a bit in some countries in some years, falling slightly in others. The same appears to be true for debt crisis governments, including governments describing themselves as implementing adjustment policies or austerity programs.

The volume of expenditure on any program cannot be taken as a measure of its impact. As already noted, within the educational sector alone, misallocation of resources in low-income countries appears to be common. Moreover, the material well-being of vulnerable groups may be directly affected by public expenditures outside these two sectors—on disaster relief, for example, or on public order or on the extension and improvement of roads. There may be less significance than is commonly supposed in either absolute levels or proportionate shares of government spending on health and education. Be that as it may: the proportion of state spending in these areas did not generally fall after the onset of the debt crisis. Furthermore the typical constraint upon spending in these sectors during the 1980s apparently was not declining availability of discretionary government resources but rather lack of interest in improving the priority of health and education expenditures on the part of governments in low-income regions.

As is well known, the 1980s was a time of slow growth and even recession in much of the third world. In and of itself this deceleration

TABLE 10–7

ESTIMATED ALLOCATION OF CENTRAL GOVERNMENT EXPENDITURES TO HEALTH AND EDUCATION FOR SELECTED GROUPS AND COUNTRIES, 1980–1988

(percent)

Group or Country	*1980*	*1982*	*1985*	*1988 (or latest year)*
Developing countries	14.7	13.5	14.0	14.5[a]
Africa	NA	NA	NA	NA
Asia	11.7	12.2	12.1	12.3
Middle East	20.6	16.5	19.4	20.9[b]
Western Hemisphere	14.8	14.8	14.0	16.0[a]
Non-oil developing countries	13.9	12.7	12.1	12.8
Major debt-rescheduling states				
Ivory Coast	20.3	NA	NA	NA
Nigeria	NA	NA	10.8	3.6[a]
Zambia	17.4	23.6	18.1	13.1[a]
Philippines	17.5	24.0	26.1	20.2[a]
Argentina	10.5	7.8	7.3	9.0[a]
Brazil	10.0	12.3	9.4	14.3[a]
Chile	21.9	21.6	19.3	18.3[a]
Mexico	20.2	14.4	10.4	8.5[a]

NOTES: Regions classified by IMF.
NA = not available.
a. 1987.
b. 1986.
SOURCE: International Monetary Fund, *Government Finance Statistics Yearbook 1989*, vol. 13 (Washington, D.C.: IMF, 1989), pp. 60–61; *1985*, vol. 9 (Washington, D.C.: IMF, 1985), pp. 48–49.

necessarily brought about adjustments in local economies. The broad pattern of such adjustments, however, appears to have been rather different from what is commonly assumed.

The UN Statistical Office prepares detailed annual estimates on the output and structure of the world's economies. These estimates currently cover the early and mid-1980s, up to the year 1986. These figures include estimates for total consumption of goods and services—a matter quite closely related to living standards (table 10–8). By those estimates consumption in poor non-Communist countries (developing

TABLE 10–8

ESTIMATED ANNUAL GROWTH IN GOVERNMENT AND PRIVATE FINAL CONSUMPTION EXPENDITURES FOR SELECTED REGIONS, 1980–1986
(percent)

Region or Country	*Government Final Consumption Expenditures*	*Private Final Consumption Expenditures*
Developed market economies	2.6	2.7
Developing market economies	2.9	2.3
Africa	2.7	1.6
Caribbean, Latin America	1.4	0.8
Asia, Middle East	2.8	3.8
Asia, East and Southeast (excluding Japan)	5.7	4.7

NOTE: Regions classified by source.
SOURCE: United Nations, *National Accounts Statistics: Analysis of Main Aggregates, 1986* (New York: UN, 1989), pp. 244–54.

market economies) rose at a slightly slower pace during the early and mid-1980s than in the industrialized West (developed market economies). The more rapid rate of population growth in the third world meant that measured per capita consumption for the low-income regions as a whole went basically unchanged. In the West, partly as a result of pervasive privatization efforts, consumption by the government sector did not rise quite as quickly as consumption by the private sector. Private individuals and nongovernment organizations appeared to gain a slightly greater say in the ultimate allocation and selection of goods and services consumed in those societies during those years.

In poorer regions, conversely, consumption by government rose at a distinctly more rapid pace than private consumption. Because growth of government consumption was kept at a pace well above the rate of population growth, per capita private consumption had to fall. In the regions hit hardest by economic downturns, the effect upon the private sector was particularly pronounced. In the impoverished sub-Sahara, by UN estimates, government consumption almost managed to keep up with the region's rapid rate of population growth, but private consumption per capita appears to have dropped nearly 10 percent. In Latin America and the Caribbean per capita consumption by government fell an estimated 6 percent between 1980 and 1986, but its fall was cushioned by private households, for whom per capita consumption is thought to have dropped about 9 percent.

Generally speaking, it would appear that government policies in low-income countries allowed the burden of adjustment to fall directly and disproportionately upon the budgets of families and private individuals under their authority. To the extent that austerity policies may be said to have affected distribution of resources during these debt crisis years, austerity apparently was imposed less on the government than on the civilian sector. Specific exceptions notwithstanding, low-income governments generally seemed to react to slower growth in the 1980s by attempting to protect state spending programs, even at the expense of private consumption by local households.

In hard times, such a course of action might seem defensible if private households could not be trusted to protect their own welfare by their spending habits, or perhaps if government programs reallocated resources preferentially to vulnerable and disadvantaged groups. Neither of these conditions, however, generally obtains in low-income regions today.

A century of research on home economics in the West, and more than a generation in developing countries, attest to the economic responses of households in all strata of the income scale.[29] Of course, human beings are much more than economic machines, and economic calculations do not drive all human actions. Nevertheless a vast literature may be adduced to demonstrate that spending decisions of households typically serve to reduce risks and to improve welfare for household members. Even the poorest, it would seem, will commonly attempt to provide for the future and willingly pay for beneficial services (including schooling and medicine for children). One might even expect the consumption patterns of poor groups to be more utilitarian and practical than for affluent groups, insofar as the margin for error is slimmer and the penalties for miscalculation and misallocation are vastly greater.

As for third world government spending programs, there is little evidence of any general preference for the vulnerable, much less the disadvantaged. If attentiveness to the condition of the poor governed the flows of state funds in low-income countries today, one would not expect prevailing policies to drain the countryside (where poverty is typically more severe) and to subsidize the cities (where living standards are typically higher). Yet with few exceptions, "urban bias" has been characteristic of state finance in low-income countries since the end of World War II.[30]

Few studies have been attempted to map out the incidence of fiscal transfers within low-income countries. But those few seem to document a pronounced tendency for subsidy and benefit to accrue not to the vulnerable, but rather to the privileged: university graduates and attendees, civil servants, unionized labor, organized urban constituencies, and army and security forces. Perhaps not coincidentally, precisely such groups can typically threaten the stability of standing third world governments. Impoverished, remote villagers and helpless children generally do not pose a corresponding challenge.

The subsidies meted out by third world governments are often considerable. In addition, a recent World Bank study indicates that losses from their state-owned enterprises—which engage in everything from winemaking to jet transportation to manufacture of unsalable goods—sometimes constitute a greater share of national income, even in debt crisis countries, than defense spending does in the United States.[31]

Household consumption in poor countries is affected not only by the incidence of transfer but by the impact of economic regulations. Commonly these confer a disadvantage upon already poor groups. Overvalued exchange rates, for example, are common in low-income regions—especially in debt crisis countries. Overvaluing a local currency may make it easier for more privileged groups to obtain or to purchase preferred goods from abroad, but it correspondingly reduces the employment opportunities for the poor by making their ordinary output (so-called labor-intensive goods) less competitive in world markets.

Peru, whose approach to the adjustment is deemed commendable by UNICEF, offers a case in point. A recent study examined Peru's structure of subsidies in some detail. It deserves to be quoted at length:

> Several features of Peru's poor have important implications. . . . First, most poor households are located in rural areas; second, most subsidies that affect rural areas do not reach the rural poor; third, the rural poor often produce much of what they consume. These facts suggest that correcting many of the distortions affecting Peru's agricultural sector may not directly harm the poor. For Peru's urban poor, the case is more complicated, but the conclusions much the same. Regarding food subsidies and taxes, were all these dropped the net (immediate) effect may well be favorable as the prices of wheat and wheat products, including those related to health and education, could be changed in ways that benefit the poor and, in some cases, reduce overall public sector expenditures.[32]

Whatever one may think of the ordinary workings of the marketplace, its results are likely to be decidedly more brutal if government policy systematically restricts and reduces purchasing power among vulnerable households within the society under its authority. Some third world governments seem to have embraced just such an approach to adjustment in the 1980s.

Generally speaking, the practices of such states do not appear to have improved their creditworthiness. They may, however, have imposed additional and unnecessary risks upon disadvantaged strata in their own populations. Fortunately, low-income households generally act to minimize the adverse impact of economic shocks or disequilibriums upon their members. Similar intentions, however, are not always evident in the policies of their governments.

Third World Priorities

The essence of UNICEF's assessment is represented by a sample of citations from *The State of the World's Children 1990*:

> No one wanted the debt crisis. . . .
>
> The consequences are falling in totally disproportionate measure on those who are least responsible for the debt and have the least capacity to repay. . . .
>
> No economic theory or political ideology can justify even a temporary sacrifice of children's growing minds and bodies. . . .
>
> Further debt reduction by Western governments (particularly for Africa) and further writing down of loans by Western banks (particularly for Latin America) is the only direction which leads anywhere for anyone. . . .
>
> The fact of the matter is that significant increases in aid will have to be paid for by the taxpayers of the industrialized nations. . . .[33]

In other words: forgive third world governments their debts and provide new funds or children will be hurt.

Over the past forty years, proponents of increased concessional transfers from Western taxpayers to third world governments have offered a variety of arguments in their proposals. A generation ago, the argument was often that such transfers could help accelerate development by contributing to state planning efforts. Fifteen years ago, one often heard that such transfers could help finance a new, more equitable international economic order in which the power of the third world states would be appropriately augmented. Today's version may not be substantially different from earlier variants, but it is far more compelling. Unlike many of the governments presiding over them, the third world's children are manifestly innocent. Furthermore, unlike the governments presiding over them, these children cannot plausibly be held responsible for the circumstances in which they find themselves.

Despite its emotional force, however, the vision of children being directly endangered by repayment of their governments' debts is fundamentally inconsistent with much of what we know about trends in the third world in the 1980s. Debt crisis notwithstanding, financial resources from the West continue to flow to third world governments, and terms of lending to these governments generally remain concessional. Debt crisis notwithstanding, no reliable statistical evidence to date suggests that the general postwar trends in health and educational progress have been arrested, much less reversed, in any broad region of the third world. Debt crisis notwithstanding, the role of the state in third world economies appears to have increased, not reduced, during an era of widely proclaimed adjustment and austerity.

Such systematic misreporting of events is not only unfortunate but ironic. For there does appear to be a connection between the debt crisis and needless distress on the part of children in the third world. The link is to be found in the attitude and practices of third world governments.

While the debt crisis may have been the result of many factors, a pervasive and propelling element was the extravagant and unproductive use of borrowed international funds by a large number of governments in developing areas. It is significant and telling that delinquencies and reschedulings on a grand scale commenced only after that time in the early 1980s when real interest rates in the international marketplace once again became positive. From then on, at least a portion of third world borrowings required productive applications if they were to finance their own repayment.

The outlook that intensified, and perhaps created, the financial difficulties of so many low-income countries in the early 1980s is also recognizable in the subsequent widespread resistance to genuine austerity measures—and in the evident willingness to shift the pain of the adjustment process directly onto the backs of private citizens. One may also identify it in the policies and practices that reduce consumption for poor groups, while protecting or subsidizing living standards for privileged groups during periods of economic stress.

Whatever else may be said about it, the debt crisis has helped to clarify the taxonomy of governance in the contemporary world. As it happens, the states that request and require large-scale and continuous concessional transfers from abroad to maintain their operation seem to be among the ones least averse to the prospect of exposing children (and other vulnerable groups under their administration) during periods of financial difficulty. If these governments are beset by crisis, it is only incidentally a crisis of finance. The scarcity defining such behavior is not in the realm of revenue. In the final analysis, it is not so much the lack of state resources as the lack of state accountability that imperils the neediest most in low-income countries today.

Accountability is unlikely to be encouraged by forgiving governments loans they have already obtained and spent or by arranging new, emergency transfusions of concessional capital to underwrite otherwise unsustainable programs and practices. Nor are the prospects for the children of low-income countries likely to be improved by those advocates who would harness their plight to revitalize the now aging movement for massive, unconditional global transfers.

11

World Population Trends and National Security

For better or worse, ours is a time of rapid and pronounced demographic change—at least by comparison with any earlier period in history. For hundreds, if not thousands, of years before the Industrial Revolution, the pace of global population change was negligible. (Rapid population change was characteristic principally of communities visited by catastrophe.) Between the years 1000 and 1750, by some estimates, the human population grew roughly 14 percent per century.[1] At current estimated rates the same proportionate growth is achieved in less than eight years.[2] Over the past few generations, demographic change has not only radically altered human numbers in the aggregate but has in addition profoundly affected their composition and international distribution. Further changes are expected in the generations immediately to come; one may anticipate these to be consequential as well. Little wonder that students of world affairs should wish to harness demography—the study of population change—in their quest to understand the vicissitudes of state power and the mysteries of national security.

To an extent that may today be only poorly recalled, the systematic study of population change was itself originally prompted by national security concerns. Although its roots come from the ancient Greek, "demography" actually entered the international vocabulary as a French word (*la demographie*). The *Oxford English Dictionary* traces the first use of the term to 1878, when it appeared as the title of a new journal published in Paris.[3] At the time, policy circles in France were almost obsessed with the population issue. Only a few years earlier, France had been badly beaten in the Franco-Prussian War; it then faced a united and increasingly assertive German nation. Many in France attributed its defeat and apparent decline at least in part to the population factor. Thanks to a persistent differential in fertility, Germany's population had finally surpassed France's around 1850.[4] The resultant gap between German and French populations was taken by many in France to have significant, perhaps even decisive, strategic implications. (That opinion was shared by some of France's opponents; no less a figure than Count von Moltke, chief of the general staff of the victorious Prussian army, was quoted as remarking that France's low birthrate cost her a battle every day.)[5] Looking toward the future, French patriots formed a National Alliance for the Growth of Population, which for a time sponsored much of the country's demographic research.

If such popular interest in the political significance of human numbers may have been inspired by a narrow nationalism, it was informed by a rough reading of broad trends in world affairs. The rise and ultimate, if fleeting, predominance of the European imperial order may have been powered by the Industrial Revolution, but it was accompanied by a demographic revolution as well. The latter revolution, though less remarked upon in history books, had tremendous political consequence. As Nobel economics laureate Simon Kuznets has observed,

> the acceleration of the rate of [world] population growth appeared earliest in Europe, where the influence of modern economic growth was first felt; reinforced by the much higher rates of increase on the continents to which European growth spread, the rate of growth of total population in areas of European settlement accelerated even more.[6]

The balance of world population underwent a major shift. Whereas areas of European population are believed to have accounted for about one-fifth of the world's total around 1750, by 1930 they accounted for about one-third.[7] In effect, the ratio of non-European to European populations, both loosely defined, dropped from about 4:1 to about 2:1 during this period. At their zenith European powers not only bestrode the earth, they also peopled it.

The rise of European population, however, did not mean enhanced political power or national security for all peoples in Europe. The state of Poland, for example, was liquidated by its neighbors at the end of the eighteenth century, and the Polish populace, despite relatively rapid rates of natural increase, spent the entire nineteenth century as subjects of the Russian, Prussian, or Austro-Hungarian empire. Conversely, while French patriots failed in their quest to raise national birthrates or to stem the relative decline of their country's demographic standing, their fears of impending political irrelevance proved unfounded. During the past century, for all its anxiety about *depopulation,* France has been on the winning side against Germany in two world wars, obtained one of the five permanent seats on the United Nations Security Council, developed a *force de frappe* (Europe's most powerful nuclear arsenal), and finally surpassed its old rival, Great Britain, in per capita economic output.

These reminders point to a paradox. While the role of population in world affairs may seem to be intuitively self-evident, its actual pertinence to the specific assessment of state power and national security is often far from obvious. Power and security are highly complex quantities. They can be described in broad generalizations, but they are tested (and thus defined) by immediate and limiting considerations. Demographic change is but one of many possible considerations limiting a state's ability to impose its will abroad or to maintain itself at home. It need not be a major factor—nor should one presume that demographic forces typically pull together to exert a single pressure on a society or state when they do come to bear.

Even when the population factor stands to have considerable influence on international affairs, the discipline of demography may provide only limited insights and guidance. Because demography deals with real people and complex mathematical formulas (often brought together by imposing computer programs), the expectant and the hopeful may accord it an unwarranted authority.

When explanations of population change are attempted, ambiguities, deficiencies, and inconsistencies typically remain. As historian Charles Tilly has noted about fertility change, "The problem is that we have too many explanations which are individually plausible in general terms, which contradict each other to some degree, and which fail to fit some significant part of the facts."[8]

As for population predictions, these have persistently disappointed users expecting precision. In 1931 Louis Dublin, then among the foremost demographers in the United States, made the forecast—on assumptions he warned were "altogether too optimistic"—that the U.S. population would peak by 1990 at 154 million persons and decline thereafter.[9] His effort resulted in an underestimate of our current population by about 100 million persons, in an estimate of total population growth between his day and our own roughly four times too low, and even in an incorrect indication of the direction of natural increase for our time. Yet Dublin's forecasts for the United States were not considered more fantastic than those of his contemporaries, nor

TABLE 11–1
CHANGING PROJECTIONS FOR POPULATION OF SELECTED REGIONS IN THE YEAR 2000
(millions)

Region	*1963*	*1988*	*Difference (%)*
Africa	768	872	+14
Latin America	638	540	–15
North America	354	295	–17

NOTE: "Africa" is defined to include North Africa.
SOURCE: United Nations, *World Population Prospects 1988* (New York: UN Department of International Economic and Social Affairs, 1989), p. 41.

have subsequent projections proved markedly more free of error. Since the 1950s the United Nations has regularly published population projections for the various regions and countries of the world. As table 11–1 demonstrates, periodic reassessments have often revised these projections rather substantially: sometimes upward, sometimes downward. In point of fact the 1963 UN projections missed the actual 1980 U.S. population by a slightly greater margin than had Dublin's forecast a similar number of years after publication. Until specialists can forecast tastes, attitudes, and capabilities precisely and far into the future, long-range population predictions are likely to prove accurate only by accident.

These limitations recognized, careful examination and reasonable use of population data may nonetheless shed at least a bit of light on some of the problems facing a few of the participants in the global arena today. Although these indications for the most part are fairly elementary, they may draw attention to neglected weaknesses or may help correct erroneous impressions that would prove costly if embodied in policy.

A brief survey can only touch upon topics deserving a more thorough treatment. This survey touches upon three general topics. It first addresses some common misconceptions about the current impact of population change on national and international security. Then it describes a few of the specific ways in which the various major components of national population change—fertility, mortality, and migration—may affect the power and security of certain states today. It concludes with some speculative remarks about the relationship between population change and Western security in the decades ahead.

Growth in Low-Income Countries

Perhaps no demographic trend in the postwar era has occasioned more commentary or aroused as much public concern as the acceleration of population growth in low-income countries. Between 1950 and 1985, according to the most recent estimates of the UN Population Division, the population of the more-developed regions (defined to include all of Europe and the USSR, Japan, North America, and Oceania) grew about 41 percent. During that same period the population of the less-developed regions—the rest of the world—is estimated to have risen about 119 percent, or almost three times as much.[10] Although the tempo of growth varied by country and by year, this population explosion has affected virtually every society in Africa, Asia, and Latin America. While fertility has been dropping for these regions as a whole—according to UN estimates the total fertility rate fell about two children per woman between the early 1950s and the early 1980s[11]—the total population of these areas was thought to be growing by more than 80 million persons each year in the late 1980s, and the actual pace of growth may actually have risen slightly in recent years.

Within intellectual and policy circles in Western countries, rapid population growth in the third world is commonly viewed as a serious problem, sometimes as a pressing one. The reasoning underlying such judgments may broadly be termed Malthusian (irrespective of whether Malthus himself would have endorsed the views and recommendations of his present-day disciples). Rapid growth of poor populations, it is argued, can mean only the spread of poverty. Population growth moreover is envisioned as eating away at economic growth in poor countries, reducing or altogether canceling potential improvements in living standards, and aggravating such conditions as poor health, malnutrition, illiteracy, and unemployment.

The political implications of such trends are often held to be ominous. By some assessments rapid population growth threatens to destabilize governments in low-income countries—by causing food shortages, for example, or by overwhelming the state with social service demands, or by creating an unmanageable and volatile crush in urban areas.[12] By contributing to a widening gap between rich and poor countries, it is sometimes further suggested, rapid population growth increases the risks of a general confrontation between haves and have-nots in the world arena. Finally, it is sometimes said, by creating major new demands for global resource use, rapid population growth in low-income countries pushes all of mankind toward an era of scarcity, perhaps even toward an unsustainable overshoot of our environment's carrying capacity.

The U.S. government takes such arguments seriously enough to have established a special population office within its foreign aid agency (USAID), administratively separate from its health program area and specifically charged with promoting lower birthrates in recipient countries.[13] Explicitly antinatal family planning activities in low-income countries are now sponsored and funded by such international organizations as the World Bank and the United Nations Fund for Population (UNFPA) and through additional, bilateral initiatives by many Western governments. Thanks in large part to such efforts, the governments of certain populous low-income countries, such as Bangladesh, had larger budgets by the early 1980s for family planning and population programs than for all other health services combined.[14]

The vision of a population explosion consuming the world as we know it conjures up enormously powerful emotions. When the passions of faith are aroused, it is exceedingly difficult to persuade the convinced by deductive reasoning or empirical evidence. Yet considerable empirical evidence suggests that the consequences and implications of the contemporary pattern of rapid population growth are significantly different from those commonly supposed.

Take the notion that rapid population growth has prevented economic progress in low-income countries. Angus Maddison, long the director of the Development Research Center at the Organization for Economic Cooperation and Development (OECD), has recently produced estimates of economic growth rates over the twentieth century for a sample of thirty-two countries, whose populations include about three-fourths of the current estimated world total (table 11–2). His sample includes such places as China, India, Indonesia, Brazil, Bangladesh, Pakistan, and Mexico, although for want of reliable data it excludes all of sub-Saharan Africa. By Maddison's reckoning, per capita output for his sample rose by a factor of more than four between 1900 and 1987. Although the populations of his nine Asian countries more than tripled during this period and the population of the six Latin American countries he covered rose by a factor of nearly seven, per capita output is estimated to have risen dramatically as well—by a factor of more than three for the Asian group and by a factor of nearly five for Latin America. Moreover, despite Latin America's highly publicized economic problems in the 1980s, per capita income in the Latin American countries in Maddison's sample—countries accounting

TABLE 11–2
ESTIMATED TOTAL AND PER CAPITA OUTPUT AND GROWTH FOR SELECTED AREAS, 1900–1987

	OECD Countries (16)	*Asian Countries (9)*	*Latin American Countries (6)*	*USSR*	*Total Countries (32)*
GDP in billion international dollars at 1980 prices					
1900	603	304	30	98	1,035
1987	7,759	3,203	982	1,684	13,629
ratio (1900 = 1)	12.9	10.5	32.4	17.2	13.2
Per capita GDP in international dollars at 1980 prices					
1900	1,946	405	645	797	841
1987	11,073	1,332	3,107	5,948	3,678
ratio (1900 = 1)	5.7	3.3	4.8	7.5	4.4
Rate of growth of per capita GDP (annual average compound rate)					
1900–50	1.1	–0.2	1.5	2.1	1.1
1950–87	3.3	3.5	2.2	2.2	2.5
1900–87	2.0	1.4	1.8	2.2	1.7

NOTE: These sixteen OECD countries are Australia, Austria, Belgium, Canada, Denmark, Finland, France, Germany, Italy, Japan, Netherlands, Norway, Sweden, Switzerland, United Kingdom, and the United States. These nine Asian countries are Bangladesh, China, India, Indonesia, Pakistan, Philippines, South Korea, Taiwan, and Thailand. These six Latin American countries are Argentina, Brazil, Chile, Columbia, Mexico, and Peru. USSR refers to Soviet government and prerevolutionary Russian empire. Total 1987 estimated population for thirty-two country groups placed at 3,705 million, or almost three-quarters (74 percent) of UN estimated 1987 midyear world population. GDP estimates are rounded to nearest billion dollars.
SOURCE: Angus Maddison, *The World Economy in the Twentieth Century* (Paris: Organization for Economic Cooperation and Development, 1989), p. 14.

TABLE 11–3
ESTIMATED POSTWAR HEALTH PROGRESS IN LOW-INCOME AREAS OF SELECTED REGIONS, 1950–1985

	More-Developed Regions	*Less-Developed Regions*	*Africa*
Life expectancy at birth (years)			
1950–55	65.7	41.0	38.0
1980–85	72.3	57.6	49.9
Change (years)	+6.6	+16.6	+11.9
Infant mortality (deaths per 1,000 live births)			
1950–55	56	180	187
1980–85	16	89	116
Change (%)	–71	–51	–38

NOTE: "More-developed regions" is defined to include all of Europe, North America, USSR, Japan, and Oceania. "Less-developed regions" is defined to include the remainder. "Africa" is defined to include all of sub-Saharan Africa plus Northern Africa. Decimal places in estimates as found in cited source.
SOURCE: Derived from United Nations, *World Population Prospects 1988* (New York: UN Department of International Economic and Social Affairs, 1989), p. 47.

for roughly three-quarters of the total population for Central America, South America, and the Caribbean—more than doubled between 1950 and 1987. Whatever else it may have done, rapid population growth has evidently not prevented major improvements in productivity in many of the societies most directly transformed by it.

Maddison's figures indicate that per capita growth accelerated sharply in most of the poor countries in his sample in the period after 1950—after the advent of the population explosion. Correlation does not imply causation. Yet many readers may be surprised that these two trends should have broadly coincided in the third world. Such surprise derives in part from a misperception of the causes of the recent rapid rates of population growth in low-income countries. The population explosion in low-income countries has been driven by a revolution in health (table 11–3). Between the early 1950s and the early 1980s, by current UN estimates, infant mortality rates in the less-developed regions fell by half, and life expectancy at birth rose by more than sixteen years. By this particular, hardly unimportant measure, the gap between rich and poor countries narrowed appreciably in recent decades. Even in perennially troubled Africa, health progress looks to have been substantial.

In itself, health progress signifies an improvement in living standards and may speak as well to conditions bearing upon health, such as nutrition, education, and housing. Improvements in health may also directly affect a population's economic potential. Theodore W. Schultz, Nobel laureate in economics, has drawn attention to the role of what he terms human capital in the economic process and to its importance in the sustained increase of per capita output.[15] Although human capital is difficult to measure, one may reasonably presume that its augmentation is represented in the rapid reduction of mortality and corresponding increases in overall length of life.

Human capital, to be sure, corresponds with economic potential, not actual achievement. Like other sorts of capital, it need not be utilized. It might even be depleted through injurious policies or practices. Be that as it may: the same forces that have driven the increase in population in poor countries would appear on their face to have increased the potential for widespread and continuing material advance.

Such a cautiously optimistic conclusion might seem to be challenged by the current example of sub-Saharan Africa, where troubles abound and population growth rates are thought to be the highest on earth. Many observers attribute the social, economic, and political ills of the region directly to its rapid population growth. They seldom stop to consider whether this may be a fallacy of composition. Sub-Saharan Africa does have the highest rate of natural increase of any large region in the world at the moment, and its pace apparently continues to accelerate. But sub-Saharan Africa is also currently characterized by what might be described as pervasive misrule. Tribal animosities are widespread and sometimes incorporated into government policy by the dominant group. State involvement in the local economy is often far-reaching, and more often than not mismanagement and misappropriation are the norm. Some governments have set about systematically uprooting their subjects and overturning their livelihoods, even when such groups are on the barest edge of subsistence.

Under such circumstances one would expect societies to report economic problems, entirely irrespective of any contribution population growth might make. When agricultural policies typically rob the farmer and civil disruption is common, it is not surprising that the increase of agricultural output should be problematic. When state-owned enterprises arrogate much of a country's total investment funds and then apply them to preferential subsidies or patently unproductive purposes, one may reasonably guess that economic growth will be affected. When a government enforces deliberately harsh, even brutal policies upon vulnerable villagers, one need not look to demography to explain resultant famine. Under current state policies and practices, most of the serious social and economic problems currently attributed to rapid population growth in the sub-Sahara would be expected to beset those same societies even if their population levels were entirely stationary. A "population problem" that proves in practice to be independent of a society's actual demographic conditions is a problem misdefined.

Because political power and national security resist statistical quantification in a way that certain economic and social factors do not, their relationship to rapid population growth is less easily measured and thus may necessarily be more speculative. In view of the prevailing preconceptions about the incipient dangers of demographic growth, however, it might be appropriate to counsel against excessive attention to the population factor as a source of instability in low-income countries or as an impetus toward "North-South" confrontation.

As we shall see, in some circumstances particular manifestations of population growth have contributed directly to the collapse of governments in certain third world countries. Blanket generalizations about the generic impact of rapid population growth are another matter altogether. Whatever else it may be, population growth is quite obviously a form of social change. At any given time some governments and societies seem to cope well with social change; others do not. For a government whose ability to cope is marginal to begin with, rapid population growth may present it with yet another unmanageable crisis. At least in the short run, however, population growth is a relatively predictable form of social change. By comparison with harvest fluctuations, rates of inflation, and many other phenomena, it is also a relatively slow form of change.

A global confrontation between rich and poor countries looks rather unlikely at the moment, if only because so many low-income countries seem to be preoccupied by their disagreements with neighboring third world states. Peter Bauer has offered the intriguing argument that North-South confrontation is exacerbated, and perhaps even caused, by the institution of foreign aid:

> The concept of the Third World and the policy of official aid are inseparable. Without foreign aid there is no Third World. Official aid provides the only bond joining its diverse and often antagonistic constituents. . . . Individual Third World countries are often neutral or even friendly to the West, but the organized and articulate Third World is invariably hostile. Its purpose as a collectivity is to coax or extract money from the West.[16]

Ironically, if Lord Bauer is correct, aid to reduce third world population growth may itself play a role in preserving North-South tension.

What of the concern that continued population growth will place a devastating burden upon the global environment, endangering the well-being of all? When public opinions are as strong and popular emotions as inflamed over any issue as they seem to be over global environmental degradation today, a few words are unlikely to change many minds. One may note, however, that modern man has been worrying about impending resource exhaustion and environmental catastrophe for more than a century. Today's attention to the ozone layer, the greenhouse effect, and the destruction of the Brazilian rain forest have precursors in nineteenth-century England's alarm that it would soon run out of coal (as the eminent economist Stanley Jevons had prophesied) and President Theodore Roosevelt's warning of a "timber famine"—disasters that failed to materialize. While the inaccuracy of past predictions does not invalidate current environmental concerns, it should raise questions about why such dramatic forecasts in the recent past have been so recurrently amiss.

One possible explanation is that such assessments have paid inadequate attention to the economic process that generated the demand for resources and put them to use. Between 1900 and 1987, by UN estimates, the world's population more than tripled, and its level of economic output, to generalize from Maddison's sample, may have increased more than a dozen times. Despite such growth of demand, the inflation-adjusted prices of many primary products—ore, farm goods, and the like—are lower today than at the turn of the century.[17] By the information that prices are meant to convey, many resources would appear to be less scarce today than they were at the turn of the century.

How could this be so? Quite simply: the economic process prompts responses to shortage and scarcity. To oversimplify, the price mechanism identifies scarcity through the agency of higher prices, thereby encouraging substitution and rewarding innovation within the limits of human preference. Previously worthless materials are brought into use (bauxite, petroleum); previously plentiful resources are more likely to be husbanded (German forests).

Environmental degradation is reported to be a serious problem in a number of countries today—China, the USSR, and the various states of Eastern Europe among them.[18] There is reason to believe that misuse of resources has been so severe in some of these countries as to contribute to a worsening of public health, as noted later. These territories have been governed by Marxist-Leninist regimes with planning mechanisms that deliberately ignored price signals in the national allocation and utilization of resources. Population growth in most of these territories is relatively slow today; a few of them even register slightly negative rates of natural increase today.

Much remains unknown about the workings of our global environment; considerably more may be understood about the general workings of the economic exchange process.[19] Uncertainties attendant upon our environment may indicate that we should act with prudence on environmental questions—but it is not necessarily clear in which direction prudence points. Until we better understand our surroundings, it may be unwise to ignore the possibility that forceful initiatives to "save the environment" may have a more adverse impact on human populations and existing political systems than would the actual tendencies such efforts propose to control.[20]

There is one implication of rapid population growth that has *not* occasioned much commentary, although it may bear directly upon the balance of world power and prospects for national security. The rapid growth of third world population and the attendant rise in per capita output in low-income areas have been affecting the distribution of global economic output—and not in the direction generally presumed. Maddison's sample points to the tendency. By his estimate, the sixteen OECD countries in his study—call them "industrial democracies" as a shorthand—accounted for about 68 percent of the total output for his group in 1950. By 1987 these same industrial democracies were estimated to account for only 58 percent of the group's output.[21] The declining relative potential of the countries currently defined as Western may indeed have important political and security implications, as noted in the conclusion.

Aspects of Population Change

The immediate impact of the population factor on national power and international security is best illustrated in specific examples. At the national level, population change is propelled by three demographic forces: fertility, mortality, and migration. Let us examine each of these components in turn.

Fertility as a Factor in World Affairs. The mathematics of demography can easily demonstrate fertility's tendency to dominate other demographic forces in the shaping of "closed" populations. In a world where the scope for migration is limited and where mortality levels are relatively stable, fertility can be expected to act as the decisive force driving changes in the local composition and global distribution of population. Under all but the most catastrophic circumstances, neither wartime losses nor mass movements of people have as much impact on a population's evolving size and structure as ordinary shifts and fluctuations in fertility.

On occasion, a country's absolute level of fertility may motivate a government to embark in directions otherwise unexpected. Thus in 1978, despite its professed adherence to the anti-Malthusian doctrines of Marxism, Leninism, and Mao Zedong thought, the People's Republic of China promulgated a new constitution in which the citizen's obligation to practice family planning was officially established, and began to implement a far-reaching campaign extolling a one-child norm for the country's parents. Although the campaign has progressed through varying phases of intensity, it apparently continues to the present day.[22]

In the view of China's current leadership, high levels of fertility were threatening the country's security. The group that came to power around Deng Xiaoping after Mao's death was evidently alarmed both by the weakness of the state they inherited and by the outlook for relations with their then-menacing neighbor, the Soviet Union. In 1978 the new leadership reemphasized the urgency of what had been called the "four modernizations" (agriculture, industry, technology, and defense) and declared that further population growth would only impede this initiative.

Beijing's diagnosis of China's "population problem" is somewhat idiosyncratic. While China's per capita output was apparently no higher (perhaps even a bit lower) in 1978 than it had been two decades earlier,[23] the policies of the Maoist era go far toward explaining this poor record. For more than twenty years, total factor productivity in the Chinese economy may have been in decline.[24] Reversing that decline required not only a cessation of many of the most obviously injurious practices of the Maoist era but also a relaxation of various state controls over the country's planned economy. Evidently, however, there were limits to the leadership's willingness to relax. Recent actions, in Tiananmen Square and elsewhere, emphasize the Chinese Communist party's commitment to maintaining its control over society and economy. While per capita output might be more expeditiously augmented through a further relaxation of existing controls and restrictions, radical population control measures apparently are more to the government's preference, and are arguably more consonant with the current leadership configuration's particular definition of state security.

Despite its continuing experiment with tactical liberalization, China's system might still be described as closed and relatively inflexible. But even in more open societies, absolute levels of fertility may influence the activities of the state. In Japan, for example, current economic security policies can easily be given a demographic inter-

TABLE 11–4
POPULATION AGED SIXTY-FIVE AND OLDER IN SELECTED INDUSTRIAL ECONOMIES, 1950–2025
(percent)

Country	*1950*	*1985*	*2025*	*Projected Shift (2025 minus 1985)*
Japan	4.9	10.3	22.2	11.9
Canada	7.7	10.4	21.0	10.6
United States	8.1	11.9	19.6	7.7
France	11.4	13.0	20.6	7.6
Italy	8.3	12.7	22.0	9.3
West Germany	14.0	14.7	23.9*	9.2
United Kingdom	10.7	15.1	20.1	5.0

NOTES: Projections for 2025 are medium variant as of 1988 assessment.
* = United Germany.
SOURCE: United Nations, *World Population Prospects 1988* (New York: UN Department of International Economic and Social Affairs, 1989), pp. 316, 370, 378, 416, 420, 558, 562.

pretation.

Between 1965 and 1989, Japan's balance of payments with the rest of the world was positive in all but three years. From the mid-1980s to the present, Japan's annual current account surplus has averaged almost $70 billion annually.[25] Japan's large and persisting surplus has become a point of irritation, even tension, with some of its trading partners and allies (most notably the United States).

Of today's industrialized democracies, Japan was the first to experience subreplacement fertility after World War II. Since the 1950s—for well over three decades—birth rates in Japan have been below the level required to keep the country's population from eventually declining. No other country in the modern era has experienced such low rates of fertility over such a long period. Low fertility, abetted by rapid postwar health progress, is rapidly transforming Japan's population profile. As a society, Japan is aging at a pace without immediate historical precedent.

Table 11–4 outlines the tendency. Between 1950 and 1985 Japan experienced a greater shift of its population into retirement age groups than any other major Western economy. Between 1985 and 2025, by current UN projections, the same will be true again. Current OECD projections concur: whereas all but one of OECD's twenty-four members are estimated to have had a higher proportion of elderly persons in their populations than Japan in 1980, all but three are projected to have lower proportions by 2010.[26] These projections are less speculative than many others, insofar as the elderly population of 2010 is alive and approaching middle age today.

By itself, a disproportionately rapid growth of retirees would be expected to increase the pension burden and various other transfer costs upon the economy of an industrial democracy. Many adjustments and countermeasures are possible, but a number of potential responses seem to be precluded in the Japanese context. A large-scale influx of "guest workers" from nearby countries to augment the labor force, for example, would be an unwelcome prospect for many, if not most, Japanese citizens. (To this day, for example, many fourth-generation Korean residents in Japan do not enjoy full citizenship rights.) Raising the retirement age might also prove difficult, insofar as Japan's elderly have, with affluence, revealed a distinct preference for leisure. (Between 1958 and 1988, labor force participation rates for those Japanese over sixty-five dropped from 41 percent to less than 24 percent, despite general improvements in popular health and attendant improvements in working potential.)[27] Under the circumstances, the accumulation of savings today may be one of the more acceptable means for coping with the demographic pressures of the future.

At both a household and a national level, Japan's savings rate is high by comparison with those of the United States, Canada, and virtually all of the countries of Western Europe.[28] Japan's savings rates have been comparatively high since at least the early 1960s. Savings have consequences. By arithmetic definition, a country's balance of payments position must equal the difference between its domestic investment and domestic savings. With unusually high rates of savings, the potential for a relatively developed economy to run current account surpluses is correspondingly greater.

As it happens, a portion of Japan's savings has been allocated to the accumulation of overseas assets, including debt instruments issued by the U.S. government. (Whether the particulars of this strategy prove to be judicious need not concern us here.) Puzzling and frustrating as it sometimes appears to its trading partners, the workings of Japan's international economy can perhaps be better understood when the country's fertility situation is taken into account.

While the instances of China and Japan demonstrate the possibility of adducing political significance to either high or low levels of fertility, political consequences more typically devolve from fertility differentials among social groups within a country. Commonly used though it may be, the term "nation" poorly describes many of the countries in the modern world including some of the most populous. Ethnic unity is the exception, not the rule, in much of the third world. As in the past, populations within the same political boundaries are often separated by linguistic or religious differences today. Unlike the preindustrial past, fertility differentials may also often separate such groups today. Differences in fertility are at once a reflection of broader differences in outlook and a potential source of increasing tensions.

In some places, differential fertility may have actually contributed directly to the collapse of the state. Lebanon comes to mind in this context. In 1943 an unwritten agreement later known as the National Pact stipulated that political authority be shared among the country's "confessional" or religious groups in accordance with their strength in the national population. Top ministers were to be divided in a 6:5 ratio between Christians and Muslims (including the Druze sect), corresponding to the country's 1932 population census. Thereafter, as one study observes, "because of the precarious and delicate sectarian arrangement in the body politic, the government . . . deliberately avoided conducting a comprehensive update of the 1932 Census."[29] Subsequent surveys, however, underscored a pronounced difference in Christian-Muslim fertility. In the early 1970s, for example, the Christian community was estimated to have a total fertility rate of less than four children per woman, as against an estimated total fertility rate of nearly six children per woman for the Muslim community as a whole.[30] By 1975 Lebanon is widely believed to have become a Muslim-majority country. Civil war erupted in 1975, and turmoil continues intermittently to this writing.

Lebanon is a Middle Eastern country, and as such shares in the region's problems. The establishment and growth of a permanent refugee Palestinian population in the south of the country after the late 1940s (the so-called "state within a state"), great-power ambitions evidenced by its Syrian neighbor, and evolving Israeli security measures all posed threats in varying degrees to the integrity of the Lebanese state. The Lebanese civil war in 1975 moreover was sparked by an attempted assassination—an event involving only a handful of people. Nevertheless, the Lebanese government's chances of persevering in its tribulations might have been greater if a changing population balance had not simultaneously undermined the rationale for the existing order.

Israel is another Middle Eastern country facing fertility-driven security pressures. Although vastly outnumbered by its typically hostile and often militant Arab neighbor states, Israel has succeeded in preserving, even enhancing, its security since its establishment more than four decades ago, fighting and winning three major wars in the process. For reasons of state, Israel occupied the territories of Gaza and the West Bank during the Six-Day War of 1967 and has maintained administrative control of these areas ever since. Maintaining control of these regions beyond the Green Line (pre-1967 boundaries) is still viewed as essential to Israel's security prospects by both major blocs in Israel's Knesset, or parliament. Yet the arrangement engenders its own security dilemmas.

Though above replacement, and indeed higher than rates for almost all other contemporary Western populations, the fertility level of Israel's Jews has been distinctly lower than that of Israel's Arabs. Fertility rates for Palestinians in the occupied or administered territories are higher still. In 1981, by one estimate, Jewish fertility in Israel averaged less than three children per woman, about half the rate for Israeli Muslims.[31] In the 1970s, total fertility rates were as high as almost nine children per woman in Gaza and the West Bank.[32] Within Green Line Israel, almost five-sixths of the population was Jewish in 1989;[33] within the boundaries of Greater Israel, however, the ratio was barely over three-fifths.[34] Even with renewed Jewish immigration from abroad and rapid fertility decline among Palestinians in the West Bank and Gaza—Jews could become a minority population within Greater Israel within a few decades.[35]

Israel has been constituted as both a Zionist state and a Western-style democracy. In the near future, under current administrative boundaries, Israel will be forced to sacrifice one of those two principles. Surrendering either would change the character of the state, and might have far-reaching effects upon its relationship with its Western allies. Resultant changes would likely force a redefinition of the country's security situation. The *intifada* in the administered territories illustrates the tension between Zionist political authority and large elements of the Palestinian population under its jurisdiction. Demographic trends did not create that tension, but they are likely to make it increasingly central to the definition of state power in Israel.

Lebanon was, and Israel still is, a mass democracy. But even under governments that do not accord each adult a vote, the weight of numbers can affect the state's ability to augment and deploy power. The Republic of South Africa offers a case in point. In 1951, as the laws and practices of Grand Apartheid were being formalized, South Africa's Whites[36] accounted for slightly more than one-fifth of the country's enumerated population.[37] By the early 1980s Whites accounted for less than a seventh of the population within the country's 1951 boundaries. By 2020, according to official government projections, the White population would amount to no more than a ninth of the total population, barring massive net migration of whites from abroad;[38]

adjusting the projections to 1951 borders, Whites might make up less than one-eleventh of the country's total. To date, the white minority government has dealt with the growing demographic imbalance in the Republic of South Africa in two different ways. In the 1970s and 1980s it declared impoverished rural Black homelands independent; among many other consequences this affected the ratio of Whites to Blacks within the officially defined territory of the Republic of South Africa. More recently the government has been relaxing or remanding some of the restrictions of its racial order. The Mixed Marriage Act has been abolished; influx control has been rescinded; Blacks are now allowed some property rights in White areas. At this writing, there is talk of abolishing the Population Classification Act, by which citizens are categorized into different races. (The Population Classification Act was finally abolished in 1991.) These steps and indications constitute a distinct departure from the previous corpus of postwar government policy. Demographic forces did not ordain these changes; in the face of existing and envisioned demographic pressures, however, some adjustments would probably have been necessary at some point. As previous South African policies demonstrate, such adjustments need not be humane, but present leadership in South Africa seems to be opting for a relatively humane set of adjustments.

Even more than South Africa, the Soviet Union can be seen as a tangle of demographic problems. In the USSR, perhaps more than in any other large country, trends in differential fertility bore directly upon the current government's prospects for projecting power and even for maintaining its own authority.

Although the terms "Soviet" and "Russian" were sometimes used interchangeably in casual conversation, ethnic Russians accounted for barely half (50.8 percent) of the enumerated population in the USSR's 1989 census. Like South Africa, the USSR assigned each citizen a state-determined race or ethnicity (in Soviet parlance, "nationality"). More than one hundred nationalities were recognized in the country's 1979 census. By that same census less than half of the USSR's non-Russian population reported itself to have a command of the Russian language; the proportion was actually somewhat lower than in the previous census.[39] The Russian population was separated by language from the life of other Soviet nationalities but not by language alone. Consequential fertility differences are also evident, for example, between the USSR's Russian population and its populations of Muslim heritage.

According to estimates by Anderson and Silver, Russians outnumbered "Muslims" in 1959 by a ratio of 4.6:1.[40] Between 1959 and 1989 the USSR's Russian population rose about 27 percent, but its

Muslim population grew an estimated 125 percent, reducing the ratio of Russians to "Muslims" to 2.6:1. By early 1989 the USSR's population of persons of Muslim heritage might have exceeded 55 million. If these people were counted as actual Muslims (not all students of the subject would do so), the former USSR would today contain the world's fifth largest Muslim population. The former USSR's Muslim population would presently outnumber the populations of such places as Egypt, Turkey, and Iran.[41]

Over the past generation the fertility rate of the USSR's Russian population hovered around (usually below) net replacement. The "Muslim" ethnicities, by contrast, registered rates typically between two and three times that of replacement. During the 1960s and 1970s, the fertility level of several of the larger "Muslim" groups appears to have risen. During the 1980s, fertility rates for most "Muslim" groups evidently began to drop, in some cases quite rapidly. But even with continued fertility decline, "demographic momentum"—pressure for population growth as the childbearing cohorts of tomorrow swell with the children of today—would have continued to affect the balance between "Muslims" and Russians in the USSR.

Table 11–5 illustrates the potential impact of such momentum. It is derived from recent medium variant projections by W. Ward Kingkade of the U.S. Census Bureau. These projections posit a continued and steady decline of "Muslim" fertility and ultimate convergence with the Russian rate at a point just below net replacement. They cover only the Central Asian nationalities, thereby excluding more than a quarter of the country's current population of Muslim descent. From today's vantage point these projections therefore might convey a relatively cautious impression of prospective changes in the demographic balance. Even so, their general outlines are striking.

In 1979 the USSR's Russians outnumbered its Central Asian nationalities by an estimated ratio of more than 5:1. By 1990 the ratio is estimated to have fallen under 4:1. By the turn of the century, it was projected to be 3:1. By 2025 it would have been only 2:1 in these projections. Even greater shifts were envisioned in the composition of the country's eighteen-year-old males, a cohort from which military inductees are customarily drawn. At the moment the ratio of Russians to Central Asians in this cohort was estimated to be roughly 3:1. In 2000 it would have been be down to only 2:1. (We can speak with some certainty; those future eighteen-year-olds are eight years old today.) For children under age eighteen, the ratio of Russians to Central Asians is already down to 2:1. By the turn of the century it was projected to be only 1.5:1. By 2025 Russians stand to account for less than one-third of the country's children. Posited declines in fertility

TABLE 11–5

RUSSIAN AND CENTRAL ASIAN POPULATION OF THE USSR, 1979–2050

	Year				
	1979	*1990*	*2000*	*2025*	*2050*
Total population (millions)					
All USSR	264	293	314	360	370
Russians	138	147	151	157	148
Central Asians	26	38	49	79	100
18-year-old males (thousands)					
All USSR	2,638	2,196	2,591	2,608	2,381
Russians	1,231	1,034	1,147	1,032	900
Central Asians	419	354	557	726	714
Children under 18 (millions)					
All USSR	70.4	81.9	87.6	109.5	130.8
Russians	32.9	35.8	34.7	34.1	31.9
Central Asians	12.0	18.0	23.4	37.6	47.4
Percentage of USSR Total					
Total population					
Russians	52	40	48	44	40
Central Asians	10	13	16	22	27
18-year-old males					
Russians	47	47	44	40	38
Central Asians	16	16	22	27	30
Children under 18					
Russians	47	44	40	31	24
Central Asians	17	22	27	34	36

NOTE: Projections are median variant assumptions from source. "Central Asians" is defined to include Kazakh, Uzbek, Turkmen, Kirgiz, and Tadzhik nationalities but to exclude certain other nationalities characterized by strong Islamic heritages.

SOURCE: Derived from W. Ward Kingkade, "The Future of the Soviet Population," in John Saunders and Lawrence Freedman, eds., *Demographic Change and Western Security* (forthcoming).

notwithstanding, Central Asian children in this projection would outnumber them.

Despite doctrinally stipulated equality of all nationalities, Russians were the dominant element to date within the USSR's multiethnic configuration. As in the imperial order that preceded it, Russians provided the Soviet Union with its official language and supplied the overwhelming majority of political personalities within the country's ruling circle. Fertility change would directly challenge prevailing assumptions about the administration of Soviet power. Russians no longer constituted a majority of Soviet men of military age (eighteen–twenty-five). Within a decade they would no longer form the majority of the working-age population and may by then account for less than two-fifths of the country's children. Such changes had implications for Soviet military, labor, and linguistic policies. If the Russian Republic substantially underwrites living standards in Central Asian republics, as some analysts in the USSR and the West believe,[42] these changes will have consequential budgetary implications as well. None of these incipient difficulties presume or require concerted political action on the part of "Muslim" populations. Under the best of circumstances, government in the Soviet Union is likely to become a more complex task.

In themselves the USSR's fertility differentials do not consign the country to domestic disorder, or even to a reduced international stature. As the nineteenth-century example of Prince von Metternich should emphasize, the decline of a fractious multiethnic empire can be forestalled or even reversed for decades through skillful leadership, even by a single individual. The impending shift in Soviet population composition will by definition be gradual, and is therefore unlikely to set immediate constraints on the day-to-day options of the Soviet leadership. Over time, however, it may just as surely alter the boundaries of the possible. Today (1991), when central authority in the USSR seems relatively weak, centrifugal ethnic passions have come to the fore and seem to be assuming a prominence in events currently unfolding. Such forces are likely only to be enhanced by the current momentum of differential fertility.

Mortality in World Affairs. As mentioned, the twentieth century has witnessed a revolution in health. Possibly three-quarters of the total improvement in life span in the history of our species has occurred since 1900.[43] So powerful have been the forces promoting improved health that even the advent of total war has not been capable of counterbalancing them. Despite terrible loss of life and attendant devastation, life expectancy for women and men alike was higher in France in 1920 than before World War I; higher in Spain after than before the Spanish Civil War; and higher in Japan and West Germany in 1950 than before World War II.[44] The same factors that have contributed to the health revolution have also evidently established the possibility of amazingly rapid recuperation from wartime destruction. The recent histories of West Germany, Japan, and South Korea demonstrate that the loss of significant portions of the working-age population and the wounding, debilitation, and episodic starvation of some considerable fraction of surviving cohorts do not now preclude rapid restoration of prewar levels of output or a rapid subsequent pace for material advance. Moreover, despite the severe privations its people suffered during and immediately after World War II, Japan currently has the longest expectation of life (and generally the lowest age-specific mortality rates) of any country—arguably suggesting that the Japanese today are the healthiest people on earth.

In the future, health progress might be halted by some cataclysm or catastrophe. Imaginably a plague or pestilence against which human populations could not develop immunity might strike. (Some contemporary commentators believe the AIDS epidemic to be just such an affliction.) One need not look to a hypothetical future for instances of interruption and even reversals of health progress in national populations. The former Soviet Union and Eastern Europe provide us with real-life examples today.

In the 1950s the Soviet Union enjoyed a rapid drop in mortality and a corresponding increase in life expectancy. So dramatic was this health progress that the UN Population Division estimated life expectancy to be slightly higher in the USSR than in the United States in the early 1960s;[45] before World War II the U.S. level was estimated to be about a decade and a half higher. In the mid-1960s, however, mortality reductions came to an abrupt halt, and death rates for men in certain age groups began to rise. As the 1960s and 1970s progressed, death rates registered a rise for virtually all adult cohorts, male and female alike. Mortality rates apparently even began to rise for Soviet infants. Although the immediate official reaction was to withhold data on these trends, the glasnost campaign has provided recent evidence on their scope. Between 1969–1970 and 1984–1985, for example, Soviet death rates for persons in their late forties were reported to have risen by more than a fifth; for those in their late fifties, more than a fourth. Between the mid-1960s and the mid-1980s, Soviet life expectancy at birth is now reported to have fallen almost three years and to have registered a decline for women as well as men.[46]

Although the USSR was apparently the first industrial society to suffer a general and prolonged deterioration of public health during peacetime, it is no longer unique. Similar, though less extreme, tendencies have been reported in Eastern Europe over the past generation. Between the mid-1960s and the mid-1980s, life expectancy at one year of age fell an average of slightly less than a year for the European members of the Warsaw Pact;[47] for men at thirty years of age, life expectancy dropped an average of more than two years during the same period. According to the most recent estimates of the World

Health Organization, by the late 1980s total age-standardized death rates (adjusted to the WHO's European model population) were higher for the USSR and the countries of Warsaw Pact Europe than for such places as Argentina, Chile, Mexico, and Venezuela.[48]

The proximate cause of the health reversals in the former Soviet bloc have been discussed in detail elsewhere.[49] A significant and pervasive rise in deaths attributed to cardiovascular deaths figures prominently in these trends. Smoking patterns, drinking patterns, and health care policies may all play their part. Recent evidence suggests that severe environmental problems may have played a greater role than previously appreciated. Intangible and intrinsically immeasurable factors, such as attitude and outlook, may have also conceivably been involved.

Whatever the etiological origins of these trends, their implications for state power are unmistakably adverse. Rising adult mortality rates reduce the potential size of a country's work force. Between 1977 and 1988 the U.S. Census Bureau reduced its projection for the turn-of-the-century population aged twenty-five to sixty-four in Eastern Europe by about 2 million persons, or about 3 percent. Insofar as the cohort had been born by 1977, and migration was negligible, the revision basically reflected a reassessment of the impact of health trends. With deteriorating health, moreover, the economic potential of surviving groups might be constrained. To the extent, if any, that attitude and outlook factor in the decline, far-reaching and not necessarily inconsequential problems of popular morale may be indirectly indicated.

Nothing is immutable about the USSR's and Eastern Europe's unfavorable mortality trends. To the contrary: at a time of generally improving health potential, it would seem to require special effort to prevent health progress. Evidently these states were up to the task. At the least, they have proved unwilling or incapable of embracing the sorts of policies that would have forestalled such declines.

Unnatural loss of life is not unknown in Communist states. But mass murder of citizens through terror is one thing, and lingering demographic attrition is quite another. One may wonder whether acquiescence in such long-term attrition does not in itself speak to a brittleness or decay in presiding policies—and thus directly to political prospects for the states in question.

Migration in World Affairs. Paradoxically, even as our evolving economic process has been increasing the scope for, and role of, human mobility in material advance, the demographic significance of international migration has decidedly diminished in recent generations. The Age of Exploration is finished. The territories of the globe are now divided among standing governments; virtually all of them limit the absorption of new citizens from abroad in some fashion, and many presume to regulate even the right to travel.

At the turn of the century gross emigration from Continental Europe was averaging more than 8 million persons per decade, or a rate equal to roughly 2 percent of the area's total population.[50] In absolute terms, the flow is much smaller today; in proportional terms, all the more so. Although they were typically undocumented by statistics, migrations of non-European populations occurred in the nineteenth century as well. (The settling of Malaysia and Southern Africa comes to mind in this regard.) Given the reluctance of most foreign governments to welcome them as citizens, opportunities for voluntary migration remain limited today for most third world populations. Increasingly, therefore, twentieth-century emigration has become a response to catastrophe—the *Aussiedlung* of millions of ethnic German refugees into what is currently West Germany during and after 1946 or the movement into what has become the state of Israel. Upheavals and turmoil have also given rise to a distinctly new form of migration—the long-term refugee housed for decades or more in a country not his own. Millions of such people may be found today in such places as Lebanon (Palestinians), Pakistan (Afghans), and Thailand (boat people from Vietnam). Modern flows of migration have thus often served as an unhappy barometer of instability and tensions in the emigrant's native land.

Although flows of migration have been small by comparison with world population, or even with global births and deaths, they have nevertheless been significant in particular areas. In the 1880s, for example, about 670,000 Scandinavian immigrants entered the United States; that relatively small figure accounted for roughly 7 percent of the total population of the countries of origin and perhaps an even greater portion of their working-age population.[51] The political consequences of even seemingly modest streams of migration can be profound. Between 1948 and 1967, net Jewish migration into Israel averaged under 50,000 persons per year; that inflow, however, was consonant with the emergence of Israel as the region's major military power.

For a country accepting migrants, national security may be affected greatly by the manner in which the state encourages newcomers to involve themselves in local economic and political life. Saudi Arabia and other Persian Gulf states have inducted a total of several million foreigners to man and to operate their oil economies; in some of these places mercenaries from abroad even flesh out the security forces. Evidencing as they do little interest in (or perhaps capacity for) incorporating guest workers of Palestinian, Pakistani, or Korean extraction into the local social fabric, the governments of these countries must engage in a complex balancing game to ensure that national power is augmented more by the presence of these foreigners than domestic stability is compromised.

By contrast, the United States has had a markedly different approach to immigrants. Qualifications and exceptions noted (most importantly, the enormous exception of slavery), the traditional U.S. attitude toward immigrants may be described as universalistic, predicated on the assumption that one can become American by coming to share a particular set of political, social, and economic values. (Note that the universalistic presumption is not invalidated by current official U.S. restrictions on numbers of new immigrants.) Without ignoring the problems manifest in the assimilation of its newcomers, one may still judge the process to have been remarkably successful. Since the founding of the Republic, more than 50 million persons have voluntarily immigrated to the United States. Though initially a product of an English-speaking population and English political theory, the U.S. system proved capable of absorbing large numbers of persons from Ireland, from German cultures, and from successively more remote Southern and Eastern European cultures. More recently, non-European groups have figured prominently in the flow of persons adopting a new American identity. To no small degree, U.S. international power and national security today can be traced, both in a specific sense and more broadly, to its approach to immigration. For reasons intrinsic to its political order, migration has on balance dramatically enhanced U.S. security. Immigration has made it possible for the United States to transform itself from a small political experiment into a superpower.

The inability to evince migration or mobility, for its part, can constrain power and limit security of standing governments. Once again, the USSR can be used to make the point. Regional development figured prominently in the USSR's successive five-year plans. For decades Soviet planners have attempted to move labor into Western Siberia and the Soviet Far East, areas rich in exploitable natural resources. In the 1980s, however, Soviet planners could no longer take surplus laborers for granted. Soviet rates of labor force participation were unusually high (regardless of how well employees actually worked). Moreover, the growth of the European population of working ages was negligible. The most appreciable increment in working-age population accrued from the USSR's Central Asian nationalities—peoples of Muslim descent, as noted. Members of these nationalities, however, proved remarkably unwilling to leave their communities, even in the face of considerable nominal financial incentive. Indeed, in these populations there was even reluctance to settle in local cities, where the Russian and European presence was more evident. So pronounced was this aversion that two Central Asian republics, Tadzhikistan and Turkmenia, registered lower levels of urbanization in 1986 than in 1970.[52] As long as "Muslim" populations were unwilling to move and to work according to the state plan, state power, as constituted in the USSR, would suffer. The Soviet system had inefficient, inflexible, and particularistic characteristics intrinsic to its polity. These characteristics were not created by demographic trends, but demographic trends can sometimes make their costs greater and more immediate.

Population and the Impending Challenge

The role of the population factor in world affairs is both limited and diverse. Demography is the study of human numbers, but the human characteristics of those numbers—their individual and collective capabilities, outlooks, and actions—define events in the world of politics at any given time. Divorced from an understanding of those human beings themselves, population studies can provide little insight or guidance for statesmen, diplomats, or generals contemplating an uncertain future.

Caveats expounded and the record of past population projections noted, today's prospective trends may nevertheless highlight an impending problem of national security. This problem may be viewed as a demographic problem, but it may just as accurately be described as a moral and intellectual problem that is taking demographic form.

Regardless of their exact calibrations, virtually all current global population projections anticipate comparatively slow population growth in today's more-developed regions and comparatively rapid growth for the less-developed regions. With some variations they envision a general continuation of trends that have been in evidence since the end of World War II. (While projected trends are not immutable, their near-term alteration would necessitate major demographic changes, perhaps not pleasant ones.) If these trends do continue another generation or two, the implications for the international political order and the balance of world power could be enormous.

Current population trends are redistributing global population—away from today's industrial democracies. The significance of this tendency may first be considered when one ranks the world's countries by their size (table 11–6). State power is a complex quality. Even so, today as in the past the Great Powers have relatively large populations. (As A. F. K. Organski once noted, "A large population does not guarantee a large armed force, a mighty production machine, or a large market, but it is a prerequisite for these important means to

TABLE 11–6
MOST-POPULOUS COUNTRIES, 1950–2025
(millions)

	1950		1985		2025	
Rank	*Country*	*Population*	*Country*	*Population*	*Country*	*Population*
1	China	555	China	1,059	China	1,493
2	India	358	India	769	India	1,446
3	USSR	180	USSR	277	USSR	351
4	U.S.*	152	U.S.*	239	Nigeria	301
5	Japan*	84	Indonesia	166	U.S.*	301
6	Indonesia	80	Brazil	136	Pakistan	267
7	Brazil	53	Japan*	121	Indonesia	263
8	U.K.*	51	Pakistan	103	Brazil	246
9	W. Germany*	50	Bangladesh	101	Bangladesh	235
10	Italy*	47	Nigeria	95	Mexico	150
11	France*	42	Mexico	79	Japan*	129
12	Bangladesh	42	W. Germany*	61	Iran	122
13	Pakistan	40	Vietnam	60	Vietnam	118
14	Nigeria	33	Italy*	57	Ethiopia	112
15	Vietnam	30	U.K.*	57	Philippines	111
16	Mexico	28	France*	55	Zaire	100
17	Spain*	28	Philippines	55	Egypt	94
18	Poland	25	Thailand	51	Turkey	90
19	Philippines	21	Turkey	50	Tanzania	85
20	Turkey	21	Iran	47	Thailand	81
Total population		1,920		3,638		6,095
Industrial democracies						
Population		454		590		430
Percentage of total population		23.6		16.2		7.1

NOTES: Projections for 2025 reflect current UN median variant assumptions.
*Industrial democracy.
SOURCE: United Nations, *World Population Prospects 1988* (New York: UN Department of International Economic and Social Affairs, 1989), pp. 38, 84.

national power.")[53] In 1950, two of the top five countries by population—and seven of the top twenty—would be currently described as industrial democracies. Their combined populations accounted for nearly a quarter of this "big country" total. By 1985—thirty-five years later—today's industrial democracies accounted for only one of the top five, and six of the top twenty; they made up less than a sixth of the group's total population. In the year 2025, not one of today's industrial democracies is projected to rank among the top five, and only two—the United States and Japan—are projected to remain among the top twenty. (Even reunited Germany's projected population would be too small to be included.) In this future world, today's industrial democracies would account for less than one-fourteenth of the total population of the "big countries." Today's industrial democracies would figure prominently only within the world's population of geriatrics. By one recent U.S. Census Bureau projection, today's industrial democracies would account for eight of the top eighteen national populations of persons aged eighty or older by the year 2025.[54]

Whatever their ultimate accuracy, current UN projections for the year 2025 depict a U.S. population slightly smaller than Nigeria's, an Iranian population almost as large as Japan's, and an Ethiopian population almost twice the size of France. Today's industrial democracies would almost all be "little countries." Canada, one of the so-called Big Seven today, would by these projections have a smaller population than such places as Madagascar, Nepal, and Syria. In aggregate, the population of today's industrial democracies would account for a progressively diminishing share of the world total (table 11–7). Where they made up more than a fifth of the world's population in 1950, they were only a sixth by 1985, and prospectively stand to be only a tenth in 2025. By such projections, the total population of all of today's Western countries would be considerably smaller than that of either India or sub-Saharan Africa in 2025, and would not be much greater than those of the Latin American and Caribbean grouping.

Projected shifts in birth totals are perhaps even more striking (table 11–8). Although these projections posit a slight rise in fertility in more-developed regions and a steady drop in less-developed regions to near net-replacement levels, women in today's Western countries are projected to bear fewer children in total than mothers in the Islamic expanse from Casablanca to Tehran by the year 2020. They are projected to be bearing a third fewer children than mothers in Latin America and the Caribbean, less than half as many as mothers in India, and less than a third as many as those from sub-Saharan Africa.

By such projections, a very different world would seem today to be in the making. Naturally much about this world may be difficult to imagine today. Weapons, political arrangements, or other innovations that would now seem fanciful might be taken for granted a few decades hence. To the extent that one can impute a continuity from our world into the one projected, however, such trends speak to pressures for a

TABLE 11–7
CHANGING GLOBAL POPULATION BALANCE FOR SELECTED GROUPINGS AND COUNTRIES, 1950–2025

Grouping or Country	Population (millions)		
	1950	*1985*	*2025*
World	2,515	4,853	8,467
Current industrial democracies	552	766	862
India	358	769	1,445
Sub-Saharan Africa	181	455	1,365
Latin America	165	403	760
Northern Africa plus Western Asia	93	234	553
	% of World Total		
Current industrial democracies	22	16	10
India	14	16	17
Sub-Saharan Africa	7	9	16
Latin America	7	8	9
Northern Africa plus Western Asia	4	5	7

NOTE: Projections are for medium variant assumptions. All figures are rounded to nearest million or nearest percentage point. "Current industrialized democracies" is defined to include North America, Japan, Australia, New Zealand, Israel, and European non-Communist countries (as of 1988). "Northern Africa and Western Asia" is defined in accordance with current UN classifications but with population of Israel omitted.
SOURCE: Derived from United Nations, *World Population Prospects 1988* (New York: UN Department of International Economic and Social Affairs, 1989).

systematically diminished role and status for today's industrialized democracies. Even with relatively unfavorable assumptions about third world economic growth, the Western countries' share of global economic output could be anticipated to decline. With a generalized progressive industrialization of current low-income areas, the Western diminution would be all the more rapid. Holding current governments fixed but projecting demographic and economic growth forward, one can easily envision a world more unreceptive, and indeed more threatening, to the interests of the United States and its allies than the one we know today.

In this exercise of the imagination, it does not require much additional imagining to conjure up conditions in which the international situation would be even more menacing to the security prospects of the countries of today's Western alliance than was the cold war of the

TABLE 11–8
CHANGING DISTRIBUTION OF GLOBAL BIRTHS FOR SELECTED GROUPINGS AND COUNTRIES, 1950–2025

	Annual Live Births (millions)		
Grouping or Country	*1950–1955*	*1980–1985*	*2020–2025*
World	99	129	144
Current industrial democracies	12	10	10
India	17	25	24
Sub-Saharan Africa	9	20	35
Latin America	8	12	14
Northern Africa plus Western Asia	5	9	11
	% of World Total		
Current industrial democracies	12	8	7
India	17	20	17
Sub-Saharan Africa	10	20	24
Latin America	8	7	10
Northern Africa plus Western Asia	5	7	8

NOTE: See note, table 11–7.
SOURCE: See source, table 11–7.

past generation. Even without the rise of new blocs or alignments, one can envision the rise of a fractious, contentious, and often inhumane international order in which liberal precepts have steadily less impact on international action and Western notions about human rights prove a progressively weaker constraint upon the exercise of force in the world arena. Even without an aggressive Soviet bloc or the invention of new weapons, this could be a very dangerous place.

In our day the proximate guarantor of ultimate Western security has been U.S. force of arms, around which the various Western security alliances have been arranged. Security, however, is a matter not only of power but of the ends for and means by which power is exercised. U.S. power has been guided by a distinctive set of principles and precepts. Broadly speaking, these principles and precepts are shared by all the governments and populations in today's Western countries. Their particulars include a respect for individual rights, such as the right to private property; adherence to a genuine rule of law; an emphasis on the civil rights of the citizen; an affirmation of the propriety of limited government; and a belief in the universal relevance of these principles. These values and precepts are not shared, or only intermittently acknowledged, by the states presiding over the great majority of the world's population today. The distinction in large part defines our security problem today—and points to our security problems tomorrow.

How to increase the share of the world's population living under such Western values? Some writers have endorsed the notion of pronatalist policies for the United States and other Western countries. Imaginative as such proposals may be, their results are likely to be of little demographic consequence. To date, pronatal efforts in Europe and elsewhere have proven expensive (as might be expected when the state gets into the business of buying children for their parents) or punitive and have had only a marginal long-term impact on fertility.[55] A government reflecting the will of the people moreover is unlikely to implement measures that would actually transform popular behavior in such an intimate and important realm as family formation.

A narrow focus on pronatalism also neglects, and perhaps even undercuts, the greatest strength of Western values: their universal relevance and potential benefit for human beings. Rather than devise means to raise birthrates in societies already subscribing to these values, it might be well to think about how they can be imparted to populations for whom they are still fundamentally alien.

It is today often argued that Western values—the concept of the liberal and open order and all the notions underpinning it—are culturally specific, and therefore cannot or should not be promoted among non-European populations. Such a view, of course, is widely endorsed by governments hostile to these notions in principle, or unwilling to be constrained by them in practice. The political values of the Western order, however, are not decisively limited to populations of Western European culture and heritage. The example of Japan should demonstrate this. Specialists today argue about the degree to which Japan is an open and liberal society.[56] To the extent that Japan may be described as a liberal democracy, however, the views and arrangements predicated by Western values have apparently proved transmissible to a major non-European culture.

Speculating about the future, one may wonder if the security prospects of the Western order do not depend upon our success in seeing two, three, many Japans emerge from the present-day third world. Under current Western arrangements, Belgium does not suffer by Germany's prosperity or her power. Nor should the security of the

United States and her current allies be diminished by the economic and demographic rise of countries sharing, and defending, common political principles.

To contemplate the Japanese example, however, is to appreciate the enormity of the task. Japan's present order emerged from highly specific and arguably irreproducible conditions. Modern Japan's political system, after all, was erected under U.S. bayonets in an occupied country after unconditional surrender. Whatever else may be said about the contemporary international scene, no world wars seem to beckon us at present. Even within Europe, the transition to a liberal order remains far from complete today. The diverse soundings from Eastern Europe suggest that the prospects for such a transition are not imminent.

How to affect such a transition in current low-income regions of Latin America, Asia, and Africa? Demographers are unlikely to provide penetrating answers to the question. To contemplate the question, however, is to consider the nature of our ultimate security challenge—a challenge that we may expect to be made all the more pressing by existing and prospective demographic trends.

Notes

INTRODUCTION

1. An early history of these construction efforts can be found in John Koren, ed., *The History of Statistics, Their Development and Progress in Many Countries* (New York: Macmillan for the American Statistical Association, 1918).
2. A phrase was coined by Pitirim Sorokin in his *Fads and Foibles in Modern Sociology and Related Sciences* (Chicago: Henry Regnery Company, 1956), chaps. 7–8.
3. Cited in Leonard Schapiro, *Totalitarianism* (New York: Praeger Publishers, 1972), p. 34.
4. For an introduction to these ideas and events, see Elie Kedourie, *Nationalism*, 4th ed. (Cambridge, Mass.: Blackwell, 1993).
5. A phrase recently popularized by Francis Fukuyama, *The End of History and the Last Man* (New York: Free Press, 1992).
6. For a definitive exposition of this phenomenon, see Michael Oakeshott, "Rationalism in Politics," in his *Rationalism in Politics, and Other Essays* (New York: Basic Books Publishing Co., 1962).
7. Alan S. Milward, *War, Economy and Society, 1939–1945* (Berkeley: University of California Press, 1977), p. 76.
8. International Monetary Fund, *Government Finance Statistics Yearbook 1991* (Washington, D.C.: IMF, 1991), p. 108.
9. Ibid., p. 108.
10. Milward, *War, Economy, and Society, 1939–1945*, p. 76; *Government Finance Statistics Yearbook 1991*, p. 108.
11. The following comparisons are all drawn from *Government Finance Statistics Yearbook 1991*, p. 108. Sadly, more recent editions of this yearbook do not present summary tables of the ratio of general government expenditure to gross domestic product for member countries.
12. Angus Maddison, "Explaining the Performance of Nations, 1820–1989," in William J. Baumol, Richard R. Nelson, and Edward N. Wolff, eds., *Convergence of Productivity: Cross-National Studies and Historical Evidence* (New York: Oxford University Press, 1994), p. 22.
13. See Christine Andre and Robert J. Delorme, "The Long-Run Growth of Public Expenditure in France," *Public Finance*, vol. 33, nos. 1/2 (1978), p. 63; and *Government Finance Statistics Yearbook 1991*, p. 108. According to these calculations, government expenditures in France in 1890 amounted to about 14 percent of GDP, whereas they equaled 30 percent of India's GDP in 1987.
14. David L. Lindauer and Ann D. Velenchik, "Government Spending in Developing Countries: Trends, Causes and Consequences," *World Bank Research Observer*, vol. 7, no. 1 (January 1992), p. 63.
15. For a classic treatment of this problem, see James M. Buchanan and Gordon

Tullock, *The Calculus of Consent: Logical Foundations of Constitutional Democracy* (Ann Arbor: University of Michigan Press, 1965).

16. A daunting philosophical literature addresses this issue; as an introduction to it, one may do well to consult Friedrich A. von Hayek, "The Use of Knowledge in Society," *American Economic Review,* vol. 35, no. 4 (September 1945), pp. 519–30. An illuminating treatment of some of the practical issues involved in relying on statistics for knowledge is Oskar Morgenstern, *On the Accuracy of Economic Observations,* 2nd ed. (Princeton: Princeton University Press, 1963).
17. Figures derived from U.S. Bureau of the Census, *Statistical Abstract of the United States 1993* (Washington, D.C.: Government Printing Office, 1993), pp. 371, 471.
18. Derived from U.S. Bureau of the Census, *Poverty in the United States: 1990,* series P-60, no. 175 (August 1991), pp. 159, 195.
19. The year 1984 comes to mind, of course, because of Charles Murray's *Losing Ground: American Social Policy, 1950–1980,* which ignited a furious controversy in public policy circles that continues to this day.
20. John C. Weicher, "Mismeasuring Poverty and Progress," *Cato Journal,* vol. 6, no. 3 (Winter 1987), pp. 715–30.
21. See, for example, David M. Cutler and Lawrence F. Katz, "Rising Inequality? Changes in the Distribution of Income and Consumption in the 1980s," *American Economic Review,* vol. 82, no. 2 (May 1992), pp. 546–51, esp. pp. 548–49.
22. Looking back on her work in the early 1960s, Mollie Orshansky, the Social Security Administration researcher credited with constructing the poverty rate, noted that the initial decision to set the official poverty line at three times the cost of the U.S. Department of Agriculture's economy food plan "endowed an arbitrary judgement with a quasi-scientific rationale it otherwise did not have." (Mollie Orshansky, "Commentary: The Poverty Measure," *Social Security Bulletin,* vol. 51, no. 10 [October 1988], p. 23.) Her candid observation is broadly pertinent to the entire project she directed.
23. For the first detailed presentation of this index and the method behind it, see Mollie Orshansky, "Counting the Poor: Another Look at the Poverty Profile," *Social Security Bulletin,* vol. 28, no. 1 (January 1965), pp. 3–29.
24. The government's most comprehensive examination of the poverty rate to date was an eighteen-volume report prepared for the U.S. Department of Health, Education, and Welfare and released in 1976 under the title *The Measure of Poverty.* In the summary report of nearly 200 pages, the merits of a consumption-based measure of poverty are dealt with in a single paragraph.
25. For the details, see Daniel T. Slesnick, "Gaining Ground: Poverty in the Postwar United States," *Journal of Political Economy,* vol. 101, no. 1 (February 1993), pp. 1–38.
26. Data in the following sections of this paragraph are drawn from *Poverty in the United States: 1990,* pp. 45–50.
27. See, for example, Trude Bennett, "Marital Status and Infant Health Outcomes," *Social Science and Medicine,* vol. 35, no. 9 (November 1992), pp. 1179-87.

28. Kenneth Prewitt, "Counting and Accountability: Numbers, the Social Sciences, and Democracy," *Social Science Research Council Annual Report, 1982–83*, pp. 13–27.
29. Friedrich A. von Hayek et al., *Collectivist Economic Planning* (London: G. Routledge and Sons, 1935). Von Mises had first laid out his theoretical critique more than a decade earlier, in his treatise *Socialism*.
30. Oskar Lange, "On the Economic Theory of Socialism," *Review of Economic Studies*, vol. 4, nos. 1 and 2 (1936–37).
31. For a comprehensive diagnosis, see Janos Kornai, *The Socialist System: The Political Economy of Communism* (Princeton: Princeton University Press, 1992).
32. For an overview, see Seymour Martin Lipset and Georgi Bence, "Anticipations of the Failure of Communism," *Theory and Society*, vol. 23 (1994).
33. For an elaboration on these arguments, see Nicholas Eberstadt, "Mortality and the Fate of the Communist States," *Communist Economies and Economic Transformation*, vol. 5, no. 4 (1993), pp. 499–516.
34. Central Intelligence Agency, *Handbook of Economic Statistics, 1988* (Washington, D.C.: National Technical Information Service, 1988), pp. 24–25.
35. Or at least apparent save to those for whom defense of those numbers remains a matter of personal honor.
36. For a comprehensive treatment, see Lorenz Krueger, Lorraine J. Dalston, and Michael Heidelberger, eds., *The Probabilistic Revolution*, vols. 1 and 2 (Cambridge, Mass.: MIT Press, 1987).

CHAPTER 1: POVERTY IN MODERN AMERICA

1. Joseph Schumpeter spoke to a related measurement problem in *Capitalism, Socialism, and Democracy* (New York: Harper & Row, 1962), p. 67:

   > There are no doubt some things available to the modern workman that Louis XIV himself would have been delighted to have yet was unable to have—modern dentistry for instance. On the whole, however, a budget at that level had little that really mattered to gain from capitalistic achievement. . . . Electric lighting is no great boon to anyone who has money enough to buy a significant number of candles and to pay servants to attend them. It is the cheap cloth, the cheap cotton and rayon fabrics, boots, motorcars and so on that are the typical achievement of capitalist production, and not as a rule improvements that would mean much to the rich man.

2. National Center for Health Statistics, *Cost of Illness, United States, 1980*, Series C, Analytical Report No. 3 (Washington, D.C.: U.S. Department of Health and Human Services, Public Health Service, April 1986).
3. Public Health Service, *Health, United States, 1985* (Washington, D.C.: U.S. Department of Health and Human Services, PHS, 1985).
4. Institute of Medicine, *Preventing Low Birth Weight* (Washington, D.C.: National Academy Press, 1985).

5. Joyce E. Allen and Kenneth E. Gadson, *Nutrient Consumption Patterns of Low-Income Households*, U.S. Department of Agriculture, Economic Research Service Technical Bulletin No. 1685, June 1983; David M. Smallwood and James R. Blaylock, "Analysis of Food Stamp Program Participation and Food Expenditures," *Western Journal of Agricultural Economics*, vol. 10, no. 1 (July 1985).
6. U.S. Department of Agriculture, *Money Value of Food Used by Households in the United States, Spring 1977* (Washington, D.C.: USDA Human Nutrition Research Center, August 1979); USDA, *Food Consumption and Dietary Levels of Low-Income Households, November 1977–March 1978* (Washington, D.C.: USDA Human Nutrition Research Center, July 1981).
7. Jeffrey S. Passell and J. Gregory Robinson, "Revised Demographic Estimates of the Coverage of the Population by Age, Sex, and Race in the 1980 Census" (unpublished paper, U.S. Census Bureau, Washington, D.C., April 8, 1985).

CHAPTER 2: U.S. INFANT MORTALITY PROBLEM

1. U.S. Department of Health and Human Services, *Health, United States, 1989* (Hyattsville, Md.: National Center for Health Statistics: 1990), p. 116.
2. Angus Maddison, *The World Economy in the Twentieth Century* (Paris: Organization for Economic Cooperation and Development, 1989), p. 19.
3. Opening statement of Congressman George Miller, March 20, 1990, in U.S. Congress, House, Select Committee on Children, Youth, and Families, *Hearings on Child Health: Lessons from Developed Nations*, 101st Congress, 2d session, March 20, 1990, p. 2.
4. Ibid.
5. Maddison, *World Economy in the Twentieth Century*, p. 71.
6. Ibid., p. 69.
7. World Health Organization, *World Health Statistics* (Geneva: WHO, 1988), pp. 402–6, and (1989), pp. 388–94.
8. Department of Health and Human Services, *Health, United States, 1989*, p. 117.
9. B. J. McCarthy et al., "The Underregistration of Neonatal Deaths: Georgia 1974–77," *American Journal of Public Health*, vol. 70, no. 9 (1980).
10. *Monthly Vital Statistics Report*, vol. 39, no. 4, supp. (August 15, 1990), p. 28.
11. Based on unpublished birthweight-specific death rates, as reported by the National Center for Health Statistics, for mid-1980s.
12. Miranda Mugford, "A Comparison of Reported Differences in Definitions of Vital Events and Statistics," *World Health Statistics Quarterly*, vol. 36 (1983), p. 207.
13. Ibid.
14. Ibid., p. 205.
15. UNICEF, *State of the World's Children 1990* (New York: UNICEF, n.d.), p. 127.
16. In the 1988 survey, reported annual expenditures were 4.6 times greater than reported pretax income for households with annual incomes of less

than $5,000; for the lowest income quintile, reported household expenditures were 2.2 times reported pretax income. U.S. Department of Labor, Press release 90-96, "Consumer Expenditures in 1988," February 26, 1990, tables 1, 2.

17. Ibid., table 2; the precise fraction was 17.7 percent.
18. Derived from United Nations Statistical Commission and Economic Commission for Europe, *International Comparison of Gross Domestic Product in Europe 1985* (New York: UN, 1988), p. 44.
19. United Nations Statistical Commission and Economic Commission for Europe, *World Comparison of Purchasing Power and Real Product for 1980* (New York: UN, 1987), p. 23.
20. Maddison, *World Economy in the Twentieth Century*, p. 19.
21. Ibid., p. 17.
22. U.S. Bureau of the Census, *Poverty in the United States, 1987* (Washington, D.C.: U.S. Department of Commerce, 1989), p. 9; Department of Health and Human Services, *Health, United States, 1989*, p. 59. The estimates in table 2–5 are based upon a method that produces a slightly higher poverty rate for the United States than that officially estimated.
23. Eve Powell-Greiner, "Perinatal Mortality in the United States, 1981–85," *Monthly Vital Statistics Report*, vol. 37, no. 10, supp. (February 7, 1989), p. 2.
24. Ibid., p. 1.
25. U.S. Department of Health and Human Services, *Vital Statistics of the United States: 1987*, vol. 1 (Hyattsville, Md.: National Center for Health Statistics, 1990), table 82.
26. Joel C. Kleinman, "Infant Mortality Rates among Racial/Ethnic Minority Groups, 1983–1984," *Morbidity and Mortality Weekly Report*, vol. 39, no. SS-3 (July 1990).
27. *Morbidity and Mortality Weekly Report*, vol. 39, no. 3 (August 3, 1990), pp. 521, 523.
28. U.S. Bureau of the Census, *Poverty in the United States, 1988*, Advance Report (Washington, D.C.: U.S. Department of Commerce), table 26.
29. Ibid.
30. Deborah Dawson, "Family Structure and Children's Health and Wellbeing: Data from the 1988 National Health Interview Survey on Child Health" (Paper presented to 1990 meeting of the Population Association of America).
31. Derived from Department of Health and Human Services, *Vital Statistics of the United States: 1987*, vol. 1, table 96.
32. *Monthly Vital Statistics Report*, vol. 39, no. 4, supp. (August 15, 1990), p. 10.
33. Ibid., p. 7.
34. Ibid., p. 30.
35. In many states Native Americans are counted as "white," as are many Hispanic-Americans; Norway's ranking would be even lower but for this convention.
36. UN Statistical Commission and Economic Commission for Europe, *World Comparisons of Purchasing Power and Real Product in 1980*, p. 24.
37. Derived from U.S. Congress, House of Representatives, Committee on Ways and Means, *Background Material and Data on Programs within the Jurisdiction*

of the Committee on Ways and Means, 1989 edition (101st Congress, 1st session, March 15, 1989), pp. 560, 563; U.S. Bureau of the Census, *Statistical Abstract of the United States 1989* (Washington, D.C.: Government Printing Office, 1989), p. 132.

CHAPTER 3: HEALTH, NUTRITION, AND LITERACY UNDER COMMUNISM

1. *Great Soviet Encyclopedia* (in Russian), 1949 (Moscow). I am indebted to Dr. William Knaus of George Washington Medical Center for this reference.
2. Christopher Davis and Murray Feshbach, *Rising Infant Mortality in the USSR in the 1970s* (Washington, D.C.: Government Printing Office, 1980).
3. Crude death rate figures are from Murray Feshbach, "Health in the USSR—Organization, Trends, and Ethics" (unpublished paper, 1985).
4. *Statistical Herald* (in Russian), 1986, no. 12.
5. *National Economy of the USSR 1985* (in Russian) (Moscow: 1986).
6. W. Ward Kingkade, "Recent Trends in Soviet Adult Mortality" (unpublished paper, U.S. Bureau of the Census, April 1987).
7. Ibid.
8. United Nations, *Demographic Yearbook* (New York: UN, various years).
9. See Nick Eberstadt, "Infant Mortality in the USSR: A Comment," *Population and Development Review*, vol. 9, no. 1 (1984), for the articles and interview in question.
10. Elizabeth Fuller, "Is Armenia on the Brink of an Ecological Disaster?" USIA *Addendum* 37, September 12, 1986; cited in Murray Feshbach, "Soviet Military Health Issues" in U.S. Congress, Joint Economic Committee, *Gorbachev's Economic Plans* (100th Congress, 1st session, November 23, 1987), p. 478.
11. Vladimir G. Treml, *Alcohol in the USSR: A Statistical Study* (Durham, N.C.: Duke University Press, 1982).
12. Ibid., and Werner Lelbach, "Continental Europe," in Pauline Hall, ed., *Alcoholic Liver Disease: Pathology, Epidemiology, and Clinical Aspects* (New York: New York Academy of Sciences, 1985).
13. World Bank, *World Development Report 1986* (Washington, D.C.: Oxford University Press, 1986).
14. Ibid.
15. Council of Mutual Economic Assistance Secretariat, *Statistical Yearbook—1984* (in Russian) (Moscow: Comecon, 1984).
16. Letter from Dr. Jiri V. Kotas, Czechoslovak Federal Council-in-Exile, to the author, 1987.
17. Judith Banister, "Analysis of Recent Data on the Population of China," *Population and Development Review*, vol. 10, no. 2 (1984), pp. 241–71.
18. Ibid.
19. "Poor in Abundance," *Economist* (London), June 13, 1987, p. 54.
20. United Nations, *Demographic Yearbook 1967* (New York: UN, 1967).
21. Sergio Diaz-Briquets, *The Health Revolution in Cuba* (Austin: University of Texas Press, 1983), p. 19.

22. United Nations, *Levels and Trends of Mortality since 1950* (New York: UN, 1982), pp. 174–76.
23. U.S. Bureau of the Census, *World Population Profile: 1985* (Washington, D.C.: Government Printing Office, 1986).
24. UN, *Levels and Trends of Mortality since 1950*, p. 174.
25. Kenneth Hill, "An Evaluation of Cuban Demographic Statistics, 1930–1980," in Paula E. Hollerback and Sergio Diaz-Briquets, eds., *Fertility Determinants in Cuba* (Washington, D.C.: National Academy of Sciences, 1983), pp. 214, 216, and 218.
26. Sergio Diaz-Briquets, "How to Figure Out Cuba: Development, Ideology and Mortality," *Caribbean Review*, vol. 15, no. 2 (1986), p. 9.
27. See Nick Eberstadt, "Introduction," *The Poverty of Communism* (New Brunswick, N.J.: Transaction Publishers, 1988), p. 10.
28. Bureau of the Census, *World Population Profile: 1985.*
29. Barbara A. Anderson and Brian D. Silver, "Infant Mortality in the Soviet Union: Regional Differences and Measurement Issues," *Population and Development Review*, vol. 12, no. 4 (1986).
30. *Nhan Dan*, May 22, 1959, quoted in David W. P. Elliot, "Political Integration in North Vietnam: The Cooperativization Period," SEADAG Papers on Problems of Development in Southeast Asia (New York: Asia Society, May 1975).
31. Gerard Tongas, "Indoctrination Replaces Education," in P. J. Honey, ed., *North Vietnam Today: Profile of a Communist Satellite* (New York: Frederick A. Praeger, 1963).
32. National Academy of Sciences, *Fertility Determinants in Cuba*, p. 217.

CHAPTER 4: PUBLIC HEALTH IN EASTERN EUROPE

1. See, for example, Murray Feshbach and Stephen Rapawy, "Soviet Population and Manpower Trends," in U.S. Congress, Joint Economic Committee, *Soviet Economy in a New Perspective* (94th Congress, 2nd session, October 14, 1976).
2. For example, Christopher Davis and Murray Feshbach, *Rising Infant Mortality in the USSR in the 1970s*, series P-95, no. 70 (Washington, D.C.: U.S. Bureau of the Census, 1980).
3. For example, see Anatoliy Vishnevskiy, "Has the Ice Cracked? Demographic Processes and Social Policy" (in Russian), *Kommunist*, no. 6 (1988), translated in Joint Publications Research Service (JPRS), series UKO, no. 88-011 (July 1988).
4. Sadly, Albania must be excluded from this assessment.
5. K. Uemura, "International Trends in Cardiovascular Disease in the Elderly," *European Heart Journal*, no. 9, supp. D (1988).
6. United Nations, *World Population Prospects as Assessed in 1984* (New York: UN Department of International Economic and Social Affairs, 1986).
7. A. Klinger, *Infant Mortality in Eastern Europe, 1950–1980* (Budapest: Statistical Publishing House, 1982), p. 2.

8. Marek Okolski, "Demographic Transition in Poland: The Present Phase," *Oeconomia Polona*, no. 2 (1983), p. 211.
9. Ibid.
10. See, for example, "Birth and Death in Romania," *New York Review of Books*, October 26, 1986.
11. Adjustments computed by ascribing the same ratio of post-neonatal mortality to total infant mortality to Bulgaria, Yugoslavia, and Romania as reported for Czechoslovakia.
12. For example, R. H. Dinkel, "The Seeming Paradox of Increasing Mortality in a Highly Industrialized Nation: The Example of the Soviet Union," *Population Studies*, vol. 39, no. 1 (1985), pp. 87–97.
13. S. Horiuchi, "The Long-Term Impact of War on Mortality: Old-Age Mortality of the First World War Survivors in the Federal Republic of Germany," *Population Bulletin of the United Nations*, no. 15 (1983).
14. Zbigniew J. Brzezinski, "Mortality Indicators and Health-for-All Strategies in the WHO European Region," *World Health Statistics Quarterly*, vol. 39, no. 4 (1986), p. 365.
15. Ibid.; his citation is M. C. Kelson and R. F. Heller, "The Effect of Death Certification and Coding Practices on Observed Differences in Respiratory Disease Mortality in Eight EEC Countries," *Revue d'Epidemiologie et de Sante Publique*, vol. 31, no. 4 (1983), p. 423.
16. "WHO MONICA Project: Geographic Variation from Cardiovascular Diseases," *World Health Statistics Quarterly*, vol. 40, no. 2 (1987), p. 171.
17. G. Lamm, *The Cardiovascular Disease Programme of WHO in Europe: A Critical Review of the First Twelve Years* (Copenhagen: WHO Regional Office for Europe, 1981), p. 35.
18. WHO MONICA Project Principal Investigators, "The World Health Organization MONICA Project (Monitoring Trends and Determinants in Cardiovascular Disease): A Major International Collaboration," *Journal of Clinical Epidemiology*, vol. 41, no. 2 (1988), p. 106.
19. V. Grabauskas et al., "Risk Factors as Indicators of Ill Health," in E. I. Chazov, R. G. Oganov, and N. V. Perova, eds., *Preventive Cardiology: Proceedings of the International Conference on Preventive Cardiology: Moscow, June 23–26, 1985* (New York: Harwood Academic Publishers, 1987), p. 308.
20. For one comprehensive review of the evidence, see *Smoking and Health: A Report of the Surgeon General* (Washington, D.C.: U.S. Department of Health and Human Services, Public Health Service, 1979).
21. Ibid., chaps. 2 and 3. See also *Health Consequences of Smoking: Cardiovascular Disease: A Report of the Surgeon General* (Washington, D.C.: U.S. Department of Health and Human Services, Public Health Service, 1983), especially p. iv: "Cigarette smoking should be considered the most important of the known modifiable risk factors for coronary disease in the United States."
22. Samuel H. Preston, *Older Male Mortality and Cigarette Smoking: A Demographic Analysis* (Berkeley: University of California, Institute of International Studies, 1970).
23. World Bank, *Poland: Reform, Adjustment, and Growth*, vol. 2 (Washington,

D.C.: World Bank, 1987), p. 398. Sample size was reported as 400,000.

24. Sophia M. Miskiewicz, "Social and Economic Rights in Eastern Europe," *Survey*, vol. 29, no. 4 (1987), p. 61.
25. U.S. Bureau of the Census, *Statistical Abstract of the United States: 1988* (Washington, D.C.: U.S. Government Printing Office, 1988).
26. Derived from *The Health Consequences of Smoking: The Changing Cigarette: A Report of the Surgeon General* (Washington, D.C.: U.S. Department of Health and Human Services, Public Health Service, 1981), statistical annex.
27. R. J. W. Melia and A. V. Swan, "International Trends in Mortality Rates for Bronchitis, Emphysema and Asthma during the Period 1971–1980," *World Health Statistics Quarterly*, vol. 39, no. 2 (1986), p. 214.
28. See, for example, *Sixth Special Report to the Congress on Alcohol and Health from the Secretary of Health and Human Services, January 1987* (Washington, D.C.: U.S. Department of Health and Human Services, Public Health Service, 1987).
29. J. de Lint, "Alcohol Consumption and Liver Cirrhosis Mortality: The Netherlands, 1950–1978," *Journal of Studies on Alcohol*, vol. 42, no. 1 (1981).
30. See *Sixth Special Report.*
31. World Health Organization, *Alcohol Policies in National Health and Development Planning* (Geneva: WHO, 1985), p. 85.
32. Radio Free Europe/Radio Liberty, "Alcoholism in Eastern Europe," RAD Background Report 130, July 30, 1987.
33. *Rude Pravo*, August 24, 1985, translated in JPRS-EPS-098, September 27, 1985, pp. 64–65.
34. Radio Free Europe/Radio Liberty, "Alcoholism in Eastern Europe."
35. *Sixth Special Report*. "When the terms 'alcoholic,' 'alcohol abuse,' or 'problem drinker' are used to designate an alcohol abuser, it must be kept in mind that these terms are somewhat less than precise"; *Fifth Special Report to the U.S. Congress on Alcohol and Health from the Secretary of Health and Human Services, December 1983* (Washington, D.C.: U.S. Department of Health and Human Services, 1983), p. xii.
36. P. A. Compton, "Rising Mortality in Hungary," *Population Studies*, vol. 39, no. 1 (1985), pp. 77, 79.
37. "Alcohol Consumption, On-the-Job Drunkenness Down in 1987," *Nepszava*, April 10, 1987, translated in JPRS, series EER-87-119, p. 104.
38. M. Litmanowcz, "Consumption of Alcoholic Beverages" (in Polish), *Wiadmosci Statystyczne*, no. 5 (May 1985), translated in JPRS, series EPS-85-098 (September 27, 1985), pp. 159–63.
39. Radio Free Europe/Radio Liberty, "Alcoholism in Eastern Europe."
40. Miskiewicz, "Social and Economic Rights in Eastern Europe," p. 49.
41. Oral communication between S. Boethig and author, in Geneva, World Health Organization, August 1988.
42. See, for example, the recent compilation published through the National Institutes of Health: Adrian M. Ostfeld, Elaine Eaker, and T. J. Truss, eds., *Measuring Psychosocial Variables in Epidemiological Studies of Cardiovascular Disease: Proceedings of a Workshop* (Washington, D.C.: U.S. Department of Health and Human Services, Public Health Service, 1985).

43. A. Jablensky, "Mental Health Behavior and Cardiovascular Disease," in Chazov, Oganov, and Perova, *Preventive Cardiology*, p. 571.

CHAPTER 5: DEMOGRAPHIC FACTORS IN SOVIET POWER

1. On the concept of correlation of forces, see Vernon Aspaturian, "Soviet Global Power and the Correlation of Forces," *Problems of Communism*, May–June 1980, and R. Judson Mitchell, *Ideology of a Superpower: Contemporary Soviet Doctrine on International Relations* (Stanford, Calif.: Hoover Institute Press, 1982).
2. See U.S. Census Bureau, *World Population Profile: 1985* (Washington, D.C.: Government Printing Office, 1983).
3. Murray Feshbach, "Trends in the Soviet Muslim Population—Demographic Aspects," in U.S. Congress, Joint Economic Committee, *Soviet Economy in the 1980s: Problems and Prospects*, 97th Congress, 2nd session, December 31, 1982.
4. World Bank, *World Development Report 1986* (New York: Oxford University Press, 1986).
5. See, for example, Alain Blum, "La Transition Demographique dans les Republiques Orientales d'URSS," *Population*, vol. 42, no. 2 (1987), pp. 337–58, and W. Ward Kingkade, "Estimates and Projections of the Population of the USSR by Age and Sex for Union Republics" (U.S. Census Bureau, April 1987, unpublished paper).
6. Kingkade, "Estimates and Projections."
7. Rosemarie Crisostomo, "Language Training and Education in the USSR" (RAND Corporation, n.d., unpublished paper).
8. See Nancy Lubin, *Labor and Nationality in Soviet Central Asia* (New York: Macmillan, 1985).
9. United Nations, *Demographic Indicators of Countries* (New York: UN, 1982).
10. *Vestnik Statistiki*, no. 12 (1986), cited in W. Ward Kingkade, "Recent Trends in Soviet Adult Mortality" (U.S. Census Bureau, 1987, unpublished paper).
11. Kingkade, "Recent Trends in Soviet Adult Mortality"; World Bank, *World Development Report 1986*.
12. See, for example, Nick Eberstadt, "The Health Crisis in the USSR," *New York Review of Books*, February 19, 1981.
13. See Nick Eberstadt, "Health of an Empire," in *The Poverty of Communism* (New Brunswick, N.J.: Transaction, 1988).
14. World Bank, *World Development Report 1986*.
15. Cited in Murray Feshbach, "Soviet Military Health Issues," in U.S. Congress, Joint Economic Committee, *Gorbachev's Economic Plans*, 100th Congress, 1st session, November 23, 1987, vol. 1, p. 468.
16. On this score see Lubin, *Labor and Nationality*.

CHAPTER 6: CIA'S ASSESSMENT OF SOVIET ECONOMY

1. U.S. Central Intelligence Agency, *Handbook of Economic Statistics, 1989* (Washington, D.C.: CIA), pp. 30–33.
2. See Henry S. Rowen and Charles Wolf, Jr., *The Impoverished Superpower* (San Francisco: ICS Press, 1990).
3. See, for example, Robert Pear, "Soviet Experts Say Their Economy Is Worse than U.S. Has Estimated," *New York Times*, April 24, 1990, p. 14.
4. For example, Aleksandr Zaichenko, "United States–USSR: Personal Consumption (Some Comparisons)" (in Russian), *SSHA: Ekonomika, Politika, Ideologiya*, December 1988.
5. Igor Birman, *Secret Incomes of the Soviet State Budget* (Boston: Martinus Nijhoff, 1981).
6. More recently, some of these figures have been substantially revised. The instability of various series on the Soviet economy is an interesting topic in its own right but one that need not detain us here.
7. See, for example, Anatoliy Vishnevskiy, "Has the Ice Cracked? Demographic Processes and Social Policy" (in Russian), *Kommunist*, April 6, 1988.
8. *Handbook of Economic Statistics, 1989*, pp. 25–26.
9. Ibid., p. 31.
10. See United Nations Statistical Commission and Economic Commission for Europe, *International Comparison of Gross Domestic Product in Europe, 1985: Report of the European Comparison Programme* (New York: UN, 1988), pp. 15–17.
11. A brief explanation is provided in *Handbook of Economic Statistics, 1989*, p. 31 n.g.
12. See, for example, *Materialen zum Bericht zur Lage der Nation im geteilten Deutschland 1987* (Materials for the report on the state of the nation in divided Germany) (Bonn: Deutscher Bundestag, Drucksache 11/11, February 18, 1987), and D. Cornelsen and W. Kirner, "On Productivity Comparisons between the FRG and the GDR," *Wochenbericht des DIW*, no. 14 (1990).
13. Derived from "East and West Germany Following Monetary, Economic, and Social Union" (in German), *DIW Economic Bulletin*, vol. 27, no. 7 (September 1990).
14. Derived from U.S. Bureau of the Census, *Statistical Abstract of the United States 1960* (Washington, D.C.: Government Printing Office, 1960), p. 239, and idem., *Historical Statistics of the United States: Colonial Times to 1957* (Washington, D.C.: G.P.O., 1975), p. 142.
15. Population growth was more rapid in the USSR than in Western Europe in the 1980s. Even so, some of the estimates for the *per capita* growth were quite similar. For 1981–1985, for example, the CIA estimates real per capita growth to have averaged 1.2 percent per year for the European Community and 1.0 percent per year for the USSR (*Handbook of Economic Statistics, 1989*, p. 34).

CHAPTER 7: POVERTY IN SOUTH AFRICA

1. *Race Relations*, vol. 1, no. 1 (First Quarter, 1939).
2. Ibid., vol. 1, no. 3 (Third Quarter, 1939).
3. Cited in Gwendolyn A. Carter, *South Africa: The Politics of Inequality* (New York: Praeger, 1958), p. 258; the pamphlet in question is attributed to Fritz Steyn.
4. S. J. Terreblanche, "Political Economy and Social Welfare with an Application on South Africa" (unpublished paper, 1986), p. 23.
5. Republic of South Africa, *Population Census 1985: Geographical Distribution of the Population with a Review for 1960–1985* (Pretoria: Central Statistics Service, 1986), p. xix.
6. Ibid.
7. World Bank, *China: The Health Sector* (Washington, D.C.: The World Bank, 1984).
8. On the basis of figures published in World Bank, *World Development Report 1982* (New York: Oxford University Press, 1982), p. 110.
9. U.S. Bureau of the Census, *World Population Profile: 1985* (Washington, D.C.: U.S. Department of Commerce, 1986).
10. Republic of South Africa, *Report of the Science Committee of the President's Council on Demographic Trends in South Africa* (Cape Town: Government Printer, 1983), pp. 89–90.
11. Estimate derived from Republic of South Africa, *Births: Whites, Coloureds and Asians 1984* (Pretoria: Central Statistical Service, 1986).
12. The Soviet practice excludes from their infant mortality count some categories of high-risk births that are included by countries following the World Health Organization methodology; Western analyses have suggested that between 14 percent and 24 percent of Soviet infant deaths may go unregistered in official counts as a result.
13. L. M. Irwig and R. F. Nagle, "Childhood Mortality Rates, Infant Feeding, and Use of Health Services in Rural Transkei," *South African Medical Journal*, vol. 66 (1984), pp. 608–13.
14. United Nations, *Model Life Tables for Developing Countries* (New York: UN, 1982).
15. C. H. Wyndham and L. M. Irwig, "A Comparison of the Mortality Rates of Various Population Groups in the Republic of South Africa," *South African Medical Journal*, vol. 54 (1979), pp. 796–802.
16. Derived from Republic of South Africa, *South African Statistics 1985* (Pretoria: Central Statistical Service, 1986).
17. Derived from *South African Statistics 1985* and World Bank, *World Development Report 1982*.
18. J. Rufus Fears, ed., *Essays in the History of Liberty: Selected Writings of Lord Acton* (Indianapolis: Liberty Press, 1985), pp. 431–32.

CHAPTER 8: THE WORLD FOOD PROBLEM

1. World Bank, *World Development Report 1982* (Washington, D.C.: World Bank, 1982), p. 152; *World Development Report 1983* (Washington, D.C.: World Bank, 1983), p. 194.
2. World Bank, *World Development Report 1983*, p. 148.
3. According to figures from the International Monetary Fund, the M2 supply per capita (that is, money and quasi money) at official exchange rates amounted to about $28 in Chad in 1980 and to about $46 in Afghanistan in 1979. By way of contrast the M2 supply per capita in 1980 amounted at official exchange rates to about $206 in Thailand and to about $616 in Mexico. IMF, *International Financial Statistics* (Washington, D.C.: December 1985).
4. United Nations, Food and Agricultural Organization, *The State of Food and Agriculture 1984* (Rome: FAO, 1985), p. 10.
5. L. A. Paulino and S. S. Tseng, *A Comparative Study of FAO and USDA Data on Production, Area, and Trade of Major Food Staples* (Washington, D.C.: International Food Policy Research Institute, 1980), pp. 67–72.
6. United Nations, Food and Agricultural Organization, *The Fourth World Food Survey* (Rome: FAO, 1977), p. 53.
7. J. S. Passell and J. G. Robinson, "Revised Demographic Estimates of the Coverage of the Population by Age, Sex, and Race in the 1980 Census" (U.S. Bureau of the Census, April 8, 1985, unpublished paper), table 2.
8. M. Lipton, *Poverty, Undernutrition, and Hunger* (Washington, D.C.: World Bank, 1983), p. 10.
9. United Nations, Food and Agricultural Organization, *Population, Food Supply, and Agricultural Development* (Rome: FAO, 1975), p. 16.
10. International Food Policy Research Institute, *Food Needs of Developing Countries: Projection of Production and Consumption to 1990* (Washington, D.C.: IFPR, 1977), p. 61.
11. U.S., Presidential Commission on World Hunger, *Overcoming World Hunger* (Washington, D.C.: Government Printing Office, 1980), p. ix.
12. U.S. Department of Commerce, Bureau of the Census, *World Population 1983: Recent Demographic Estimates for the Countries and Regions of the World* (Washington, D.C.: U.S. Bureau of the Census, 1983), p. 17.
13. United Nations, *Demographic Yearbook 1973* (New York: UN, 1974), p. 82.
14. India, Cabinet Secretariat, *National Sample Survey: Twenty-third Round, July 1968–June 1969, No. 200* (Delhi: CS, 1974), p. 9. More recent rounds of the National Sample Survey have not systematically attempted to estimate margins of error in national crop production figures. For a recent explication of the rationale and the methodology by which such figures are currently produced, see India, Central Statistical Organization, *Statistical Abstract of India 1982* (New Delhi: CSO, n.d.), pp. 42–43.
15. T. J. Goering, *Tropical Root Crops and Rural Development* (Washington, D.C.: World Bank, 1979), p. 3.
16. I. B. Kravis, Alan Heston, and Robert Summers, *International Comparisons of Real Product and Purchasing Power* (Baltimore: Johns Hopkins University Press, 1978), p. 22.

17. D. McGranahan, *International Comparability of Statistics on Income Distribution* (Geneva: UNRISD, 1979), p. 2.
18. Lipton, *Poverty, Undernutrition, and Hunger*, pp. 12–14, 34.
19. H. Alderman and S. Reutlinger, *The Prevalence of Calorie Deficient Diets in Developing Countries* (Washington, D.C.: World Bank, 1980).
20. United Nations, *Levels and Trends of Mortality since 1950* (New York: UN, 1982), p. 83.
21. F. Golladay, *Health: Sector Policy Paper* (Washington, D.C.: World Bank, 1980), pp. 5, 11.
22. UN, *Levels and Trends*, p. 109.
23. R. E. Schneider, M. Shiffman, and J. Faigenblum, *American Journal of Clinical Nutrition*, vol. 31 (1978), p. 2091.
24. D. G. Johnson, *American Statistician*, vol. 28 (1974), pp. 90, 94.
25. P. B. R. Hazell, *Journal of Agricultural Economics*, vol. 36 (1985), pp. 145–59.
26. P. B. R. Hazell, *Instability in Indian Foodgrain Production* (Washington, D.C.: International Food Policy Research Institute, 1982), p. 9.
27. World Bank, *World Development Report 1986* (Washington, D.C.: World Bank, 1986), pp. 184–85.
28. World Bank, *World Tables*, 3rd ed., vols. 1, 3 (Baltimore: Johns Hopkins University Press, 1983).
29. World Bank, *Commodity Trade and Price Trends* (Washington, D.C.: World Bank, 1986), pp. 52–55.
30. P. N. Mari Bhat, S. Preston, and T. Dyson, *Vital Rates in India, 1961–1981* (Washington, D.C.: National Academy Press, 1984), pp. 126, 146.
31. The growing commercialization of the international grain trade has not reduced the volume of food grains devoted to emergency or charitable purposes. In 1984 total food aid from Western nations and international institutions totaled an estimated 10 million tons. This was three times more than had been available a decade earlier; in absolute volume it exceeded sub-Saharan Africa's total grain imports during 1984.
32. An example of a technological improvement in the information system is the development of remote sensing satellite capabilities, which have increased the accuracy of crop forecasts for areas where harvest data are less available or less reliable. A diplomatic improvement in the information system was the long-term grain agreement between the United States and the Soviet Union, which dramatically reduced the chances that the world market would be caught unaware by a sudden, massive purchase from the USSR, the nation that was the world's largest grain importer.
33. A similar scenario guided the Southeast Asian rice market. After liquidating stocks in 1972 to meet an exceptional purchase by Indonesia, the Thai government, faced with a poor domestic rice crop and urban discontent over rising rice prices, placed a ban on all rice exports. Thailand was at that time the primary supplier of rice in the world marketplace.

CHAPTER 9: INVESTMENT WITHOUT GROWTH

1. Figures derived from Organization for Economic Cooperation and Development, *Geographical Distribution of Financial Flows to Developing Countries* (Paris: OECD, various years); International Monetary Fund, *Primary Commodities: Market Developments and Outlook* (Washington, D.C.: IMF, 1988); and U.S. Department of Agriculture, *World Agricultural Supply and Demand Estimates,* (WASDE/228, Washington, D.C.: U.S. Agriculture, Economic Research Service, March 9, 1989).
2. Several assessments of this general nature have been published in recent years. They include Robert Cassen and Associates, *Does Aid Work?* (Oxford, Clarendon Press, 1986), sponsored by the World Bank and the International Monetary Fund; Anne O. Krueger and Vernon W. Ruttan, *The Development Impact of Economic Assistance to LDCs* (unpublished volume commissioned by the U.S. Agency for International Development University of Minnesota, 1983); and Organization for Economic Cooperation and Development, *Twenty-Five Years of Development Cooperation: A Review* (Paris: OECD, 1985). The singularity of these efforts is perhaps most noteworthy. OECD does produce an annual report on ODA, *Development Cooperation,* but this is concerned principally with the *volume of resources transferred* to low-income countries, not with the *impact* of these funds.
3. World Bank, *World Development Report 1987* and *1988* (New York: Oxford University Press, 1987 and 1988), statistical appendix, table 5.
4. International Monetary Fund, *International Financial Statistics Yearbook 1988* (Washington, D.C.: IMF, 1988), pp. 172–79.
5. The date is significant. Distortions apparent in 1972 cannot be ascribed to subsequent events such as the oil shocks or the third world debt crisis.
6. World Bank, *World Development Report 1988* (New York: Oxford University Press, 1988), statistical appendix, tables 1, 2, and 5.
7. U.S. Bureau of the Census, *Statistical Abstract of the United States: 1989* (Washington, D.C.: Government Printing Office, 1989), table 1089.
8. Ibid., table 1357.
9. Marvin Schwartz, "Estimates of Personal Wealth, 1982: A Second Look," *Statistics of Income Bulletin,* vol. 7, no. 4 (1988), table 1B.
10. New York Stock Exchange, *Stocks and Warrants Available for Trading* (New York: NYSE Market and Research, December 1988).
11. World Bank, *World Development Report 1989* (New York: Oxford University Press, 1989), statistical appendix, table 25, p. 212.
12. Nicholas Eberstadt, *Foreign Aid and American Purpose* (Washington, D.C.: American Enterprise Institute, 1989), p. 84.

CHAPTER 10: THE DEBT BOMB

1. World Bank, *World Debt Tables 1989–90, External Debt of Developing Countries,* vol. 1, *Analyses and Summary Tables* (Washington, D.C.: World Bank, 1989), p. 78.

2. *New York Times,* December 20, 1988.
3. United Nations Children's Fund, *The State of the World's Children 1990* (New York: Oxford University Press, n.d.), p. 1.
4. Ibid, pp. 8–9.
5. Ibid., p. 8.
6. Ibid., p. 11.
7. World Bank, *World Debt Tables,* p. 78.
8. World Bank, *World Development Report 1989* (New York: Oxford University Press, 1989), p. 18.
9. Organization for Economic Cooperation and Development, *Development Cooperation: 1988 Report* (Paris: OECD, 1988), p. 52.
10. World Bank, *World Debt Tables,* p. 10.
11. Taiwan has typically been excluded from estimates of aggregate trends for developing countries since its replacement by the People's Republic of China in the UN General Assembly in 1971.
12. OECD, *Development Cooperation: 1988 Report,* p. 52.
13. Long-term lending is defined here as maturities of one year or longer; LIBOR rate is for one-year loans.
14. In 1972 and 1973, for example, the discount rate at the U.S. Federal Reserve Bank averaged 6.0 percent. Average interest costs for developing countries were estimated at 5.0 percent during that period and about 6.4 percent for upper-middle-income borrowers. Data from International Monetary Fund, *International Financial Statistics Yearbook* (Washington, D.C.: IMF, various issues), and Organization for Economic Cooperation and Development, *Financing and External Debt of Developing countries: 1988 Survey* (Paris: OECD, 1989).
15. A borrower may feel pressed if impelled to repay funds used for unproductive purposes. When lending contains a substantial gift element, the incentives for productive use of borrowed capital may be somewhat less immediately compelling. In the late 1980s, as in the early 1980s, debtor governments on the whole could redeem their external debt obligations by earning a lower rate of return on their borrowings than a local entrepreneur, an international business, or even a Western government. Such preferential arrangements may affect the expectations of some beneficiaries.
16. UNICEF, *State of the World's Children 1990,* p. 8.
17. Ibid., p. 9.
18. Results from reasonably complete censuses can also be used to estimate child mortality—but no more frequently than censuses are conducted and without much indication of year-to-year changes for the interval examined.
19. As explained in the study's text. See United Nations, *Mortality of Children under Age Five: World Estimates and Projections, 1950–2025* (New York: UN Department of International Economic and Social Affairs, 1988), p. 25.
20. United Nations Educational, Scientific, and Cultural Organization, *UNESCO Statistical Yearbook 1989* (Paris: UNESCO, 1989).
21. UNICEF, *State of the World's Children 1990,* p. 63.
22. Ibid.

23. Ibid., p. 57.
24. Ibid., p. 13.
25. Ibid., p. 61.
26. Ibid., p. 62–63.
27. G. A. Cornia, F. Stewart, and R. Jolly, *Adjustment with a Human Face,* vol. 1 (New York: Oxford University Press, 1987), pp. 290–93.
28. World Bank, *World Debt Tables 1989–90,* table VI.6.
29. For a succinct summary and interpretation of such research, see the Nobel Prize address by Economics Laureate Theodore W. Schultz, reprinted in his *Investing in People* (Berkeley: University of California Press, 1981).
30. This theme is further developed in Michael Lipton, *Why Poor People Stay Poor* (New York: Oxford University Press, 1977), and his subsequent work.
31. World Bank, *World Development Report 1988* (New York: Oxford University Press, 1988), p. 172.
32. Paul Glewwe and Dennis de Tray, "The Poor in Latin America during Adjustment: A Case Study of Peru," World Bank Living Standards Measurement Study Working Paper 56 (Washington, D.C., 1989), summary.
33. UNICEF, *State of the World's Children 1990,* pp. 11, 55, 57.

CHAPTER 11: POPULATION AND NATIONAL SECURITY

1. Estimates by Alexander M. Carr-Saunders, cited in Simon Kuznets, *Modern Economic Growth: Rate, Structure and Spread* (New Haven, Conn.: Yale University Press, 1966), p. 35.
2. As derived from United Nations, *World Population 1988* (New York: UN Department of International Economic and Social Affairs).
3. The French demographer Michel Huber dated it somewhat earlier: according to him, the term was coined by Achille Guillard in 1855; see Michel Huber, Henri Bunle, and Fernand Boverat, *La Population de la France: Son Evolution et Ses Perspectives* (Paris: Librarie Hachette, 1947), p. 2. The point remains the same.
4. Joseph J. Spengler, *France Faces Depopulation* (Durham, N.C.: Duke University Press, 1938), p. 25.
5. Ibid., p. 131.
6. Kuznets, *Modern Economic Growth,* p. 39.
7. Derived from Carr-Saunders' estimates, in ibid., p. 38.
8. Charles Tilly, "The Historical Study of Vital Processes," in Charles Tilly, ed., *Historical Studies of Changing Fertility* (Princeton, N.J.: Princeton University Press, 1978), p. 3.
9. As quoted in William Petersen, *Population* (New York: Macmillan, 1975), p. 339.
10. Derived from UN, *World Population 1988,* p. 38.
11. Ibid., p. 204.
12. See, for example, the summary volume of National Academy of Sciences, *Rapid Population Growth: Implications and Consequences* (Baltimore: Johns Hopkins University Press, 1971).

13. Indicative of the program's founding emphasis is the introductory paragraph in the Office of Population's 1973 annual report:

> Fiscal year 1973 was a year of heightened awareness of the crisis proportions of the world population problem. Emerging demographic data continued to show that many populations were multiplying at alarming rates. The threat of rapid resources depletion became more clearly apparent. Looming energy shortages added to the growing evidence of modern man's inability to sustain himself in ever-expanding numbers.

    USAID, *Population Program Assistance: United States Aid to Developing Countries* (Washington, D.C.: Office of Population, May 1974), p. 1.
14. World Bank, *World Development Report 1984* (New York: Oxford University Press, 1984).
15. See, for example, his *Investing in People* (Berkeley: University of California Press, 1981).
16. Peter Bauer, *Reality and Rhetoric* (Berkeley: University of California Press, 1984), p. 33.
17. The tendency was most intensively examined in Harold Chandler and Barnett Morse, *Scarcity and Growth* (Washington, D.C.: Resources for the Future, 1963). For a more recent treatment, see the collection of studies in Julian Simon and Herman Kahn, eds., *The Resourceful Earth* (London: Basil Blackwell, 1984).
18. See, for example, Vaclav Smil, *The Bad Earth* (Armonk, N.Y.: M. E. Sharpe, 1983); Boris Komarov, *The Destruction of Nature: The Intensification of the Ecological Crisis in the USSR* (Frankfurt: Posev Verlag, 1978); and John Lampe, ed., *Environmental Crises in Eastern Europe* (forthcoming).
19. The laws of thermodynamics, it is today believed, ultimately constrain the closed system we take the universe to be. As for earthly constraints, however, the only nonrenewable economic resource without possible substitution is the time of human beings. The value of human time has generally and substantially increased throughout this century, despite the multiplication of human beings.
20. Affluent populations around the world have revealed a distinct preference for a cleaner environment. Even if environmental protection were to be regarded purely as a luxury, it is one that prosperous peoples are not only willing but often eager to pay for.
21. Derived from Angus Maddison, *Economic Growth in the Twentieth Century* (Paris: Organization for Economic Cooperation and Development, 1989), p. 4.
22. See, for example, John S. Aird, *Slaughter of the Innocents: Coercive Birth Control in China* (Washington, D.C.: AEI Press, 1990).
23. Nicholas R. Lardy, *Agriculture in China's Modern Economic Development* (New York: Cambridge University Press, 1983).
24. K. C. Yeh, "Macroeconomic Changes in the Chinese Economy during the Readjustment," *China Quarterly*, no. 100 (1984).
25. International Monetary Fund, *International Financial Statistics* (Washington, D.C.: IMF), various issues. Figures are in current dollars, at current exchange rates.

26. Robert P. Hagemann and Guiseppe Nicoletti, "Population Ageing: Economic Effects and Some Policy Implications for Financing Public Pensions," *OECD Economic Studies,* no. 12, Spring 1989, p. 56.
27. Derived from Japan, *Japan Statistical Yearbook* (Tokyo: Management and Planning Bureau), various issues.
28. Organization for Economic Cooperation and Development, *OECD Historical Statistics 1960–87* (Paris: OECD, 1989), pp. 69–70. Of all twenty-four OECD members, only Luxembourg exceeds Japan in estimated gross national savings rate, and only Italy in estimated net household savings rate, for the 1960–1987 period.
29. Thomas Collelo, ed., *Lebanon: A Country Study* (Washington, D.C.: Government Printing Office, 1989), p. 48.
30. Joseph Chamie, "Religious Differentials in Fertility: Lebanon, 1971," *Population Studies,* vol. 31, no. 2 (1977), pp. 365–82.
31. Michael Roof, "Detailed Statistics on the Population of Israel by Ethnic and Religious Group and Urban and Rural Residence: 1950 to 2010" (U.S. Bureau of the Census, Center for International Research, unpublished paper, September 1984), p. 3.
32. Gary S. Schiff, "The Politics of Population Policy in Israel," *Forum,* Winter 1978, p. 186.
33. Foreign Broadcast Information Service (FBIS), NES 89-223, September 29, 1989, p. 22.
34. Derived from ibid. and FBIS MEA April 11, 1986, pp. I-7/8.
35. Reinhard Wiemer, "Zionism, Demography and Emigration from Israel," *Orient,* vol. 28, no. 3 (1987).
36. Until 1991, by South African law (Population Classification Act) every South African was assigned a race by the state. The categories were White, Coloured, Asian, and Black; Blacks and Whites were the two largest enumerated groups.
37. *South African Statistics 1980* (Pretoria: Government Printer, 1980), p. I.6.
38. Ibid.
39. Barbara A. Anderson and Brian D. Silver, "Some Factors in the Linguistic and Ethnic Russification of Soviet Nationalities: Is Everyone Becoming Russian?" in Lubomyr Hajda and Marc Beissinger, eds., *The Nationality Factor in Soviet Politics and Society* (Boulder, Colo.: Westview, 1990).
40. Barbara A. Anderson and Brian D. Silver, "Demographic Sources of the Changing Ethnic Composition of the Soviet Union," *Population and Development Review,* vol. 15, no. 4 (1989), p. 623.
41. Derived from United Nations, *Demographic Yearbook 1988* (New York: UN, 1990).
42. For example, Nancy Lubin, *Labour and Nationality in Soviet Central Asia* (London: Macmillan, 1985).
43. Nick Eberstadt, *The Poverty of Communism* (New Brunswick, N.J.: Transaction Books, 1988), p. 11.
44. As verified in United Nations, *Demographic Yearbook* (New York: UN), various issues.

45. United Nations, *World Population Prospects as Assessed in 1984* (New York: UN Department of International Economic and Social Affairs, 1986).
46. For more details, see Eberstadt, *The Poverty of Communism* and chapter 3 in this book.
47. See chapter 4 in this book for more details.
48. World Health Organization, *World Health Statistics Annual 1989* (Geneva: WHO, 1989), table 12.
49. See, for example, chapters 3, 4, and 5 in this book.
50. Kuznets, *Modern Economic Growth,* p. 52.
51. Ibid., p. 55. Political evolution in Scandinavia may also have been affected by migration. The emergence of the Swedish welfare state, for instance, dates to World War I—when impoverished Swedes no longer had the option of emigrating to the United States.
52. Cited by W. Ward Kingkade, "Demographic Trends in the Soviet Union," in U.S. Congress, Joint Economic Committee, *Gorbachev's Economic Plans,* 100th Congress, 1st session, November 23, 1987, vol. 1, pp. 174–75; see also chapter 5.
53. A. F. K. Organski, *World Politics* (New York: Knopf, 1958), p. 147.
54. Barbara Bosle Torrey, Kevin Kinsella, and Cynthia M. Taeuber, *An Aging World,* U.S. Bureau of the Census Report, series P-95, no. 87, September 1987, p. 11.
55. Under communism, Romania's Ceaucescu regime implemented forceful and harsh pronatalist policies, but these were inadequate to keep the country's fertility rates from dropping below replacement in the 1980s.
56. For a skeptical view, see Karel van Wolferen, *The Enigma of Japanese Power* (New York: Knopf, 1988).

Index

Acton, J. E. E. D. (Lord), 168
Adjustment policy, 228–36
Afrikaner population, South Africa, 150–52, 168–69
Agricultural sector
  output of less-developed countries from, 201
  problems and performance of sub-Saharan Africa in, 189–92, 248
Aid to Families with Dependent Children (AFDC), 67, 69
Aird, John S., 288 n27
Alcoholic beverage consumption, 80–81, 111–14
Alderman, H., 284 n19
Allen, Joyce E., 274 n5
Anderson, Barbara, 256, 277 n29, 289 nn39, 40
Antipoverty policy, South Africa, 150–52, 168–69
  See also Deurbanization; Homelands, South Africa; Migration
Apartheid policy, South Africa, 6–7, 151
  See also Grand Apartheid
Armenia, 80
Aspaturian, Vernon, 280 n1
Austerity measures, developing countries, 228–29, 234
Australia, 50, 51

Banister, Judith, 276 n17
Bantustans, South Africa, 152, 155, 168–69
Bauer, Peter T., 248, 288 n21
Beissinger, Marc, 289 n39
Bennett, M. K., 37
Birman, Igor, 281 n5
Birthweight
  correlated with marital status, 57–60
  data on distribution of, 54–56
  as factor in infant mortality, 31–32
  factors contributing to, 56–57
Black homelands. *See* Homelands, South Africa
Blaylock, James R., 274 n5
Blum, Alain, 280 n5
Bophuthatswana, 152
Borrowing, international
  IMF estimates of new lending and repayments, 229
  by low-income countries, 212–13
  See also Concessional loans; Debt, external
Boverat, Fernand, 287 n3
Bremer, Karl, 150
Brzezinski, Zbigniew, 102, 278 nn14, 15
Budget deficit, USSR, 140–41
Bulgaria
  adult mortality rates in, 99–101
  cigarette consumption in, 110–11
  distilled spirits consumption in, 112–13
  health care in, 114–17
  infant mortality rates in, 94–96
  life expectancy in, 92–94, 96–99
Bunle, Henri, 287 n3

Caloric intake and requirements, 39–40, 175–76
Cambodia, 89
Capital flows

inaccurate measurement of, 215–16
international transfers as, 205–10
Cardiovascular disease, 102–8, 110, 118, 119
Carnegie Commission report on South Africa (1932, 1984), 150, 152
Carr-Saunders, Alexander, 287 nn1, 7
Carter, Gwendolyn, 282 n3
Cassen, Robert, and Associates, 285 n2
Castro, Fidel, 84, 86
CCP. *See* Chinese Communist party (CCP)
Centers for Disease Control, 29
Central Intelligence Agency (CIA)
comparison of estimates with DIW, 146
comparison with ICP estimates for USSR, 144–46
implications of estimates mismeasuring USSR economy, 23–24, 146–49
problems of estimates of Soviet economy by, 138–46
CES. *See* Consumer Expenditure Surveys (CES)
Chamie, Joseph, 289 n35
Chandler, Harold, 288 n17
Children
mortality rates of, 163, 178
perceived effect of adjustment policies on, 227–36
perceived effect of debt crisis on world's, 212–13, 220–28, 236–38
in poverty, 30–35, 51–53
See also Education; Health; Mortality rates, infant
China
dictatorship, 8
family planning in, 251–52
health system in, 82–83
infant mortality rates in, 82, 84
nutritional well-being in, 88–89
population control in, 82–83, 251–52
Chinese Communist party (CCP), 82
CIA. *See* Central Intelligence Agency (CIA)
Ciskei, 152, 162
Collelo, Thomas, 289 n29
Communist East-bloc countries
alcohol consumption in, 80–81, 111–14
levels of government spending (1981–1985), 11–12
life expectancy in, 78, 81, 96–99, 260
mortality rates in, 78–82, 93–95, 99–101, 260–61
political transformation, 5–6
spending for public health in, 81–82
See also Bulgaria; Czechoslovakia; Hungary; Life expectancy at birth; Mortality rates, infant; Poland; Romania; Yugoslavia
Compton, P. A., 279 n36
Concessional loans, 218–20
Consumer Expenditure Surveys (CES), 52, 73
Consumption
CIA estimates of USSR, 141–42
household, CES surveys, 52
household food, 35–38
as ratio of total output in developing countries, 201–3
UN Statistical Office estimates of world trends, 232–235
See also Personal consumption expenditure (PCE)
Cornelsen, D., 281 n12
Cornia, G. A., 287 n27
Crisostomo, Rosemarie, 280 n7
Cuba, 74, 85–86
Cultural Revolution, 90–91
Czechoslovakia
adult mortality rates in, 99–101
cigarette consumption in, 110–111
distilled spirits consumption in, 111–14
health care in, 114–18
infant mortality rates in, 94–96
life expectancy in 92–94, 104–7

Data
accuracy and reliability of low-income country, 199–205

CIA use of USSR, 23–24, 138–42
conflicting estimates in, 183–85
for developing country populations, 173–74
effect of misused, 15–25
falsification of USSR, 140–41
generation under command economy, 21
for investment in less-developed countries, 199–204
reliability of U.S., 49–51
reliability of world food and nutrition, 170-72, 181
of UN Population Division, 221–25
Davis, Christopher, 276 n2, 277 n2
Dawson, Deborah, 60, 275 n30
Death rates, 101–3, 106–7
changes in USSR of age-specific, 126
patterns in Communist East-bloc countries for, 101–7
See also Mortality rates
Debt, external
of developing countries, 212
identification of concessional segments of, 218–20
OECD estimates of financial transfers related to, 215–16, 218–20
perceived effect of crisis in repayment of, 212–13, 221–28, 236–38
perceived effect on government spending of, 231–33
perceived effect on world's children of, 212–13, 221–28, 236–38
rescheduling by low-income countries of, 228–29
rising debt-service payments for, 214–15
World Bank estimates of financial transfers related to, 216–18
See also Austerity measures; Capital flows; Concessional loans; London Interbank Offered Rates (LIBOR), rate loans; Wealth
Decolonization, 5
de Lint, J., 279 n29
Demographic trends
effect on power and security of changing, 239–41
effect in USSR of, 120, 257
See also Fertility; Labor force; Mortality rates
Deng Xiaoping, 83
Dependency indicator, 34
Deprivation
defined, 52
indicators of, 28
de Tray, Dennis, 287 n32
Deurbanization, South Africa, 166
Developing countries. *See* Less-developed countries
Development policy, less-developed countries, 202–3
Diaz-Briquets, Sergio, 276 n21, 277 n26
Dinkel, R. H., 278 n12
Disease
incidence in Communist East-bloc countries of, 102–7
incidence in Cuba of, 85
incidence in USSR of, 128
DIW. *See* German Economic Research Institute (DIW)
Dublin, Louis, 241–42
Dyson, T., 284 n30

Eaker, Elaine, 279 n42
Eastern Europe. *See* Communist East-bloc countries
East Germany. *See* German Democratic Republic (GDR)
Eberstadt, Nicholas, 276 n9; 277 n27; 280 nn12, 13; 289 n43; 290 n46
Economic assistance statistics for policy, 24–25
Economic growth, less-developed countries, 209
Economic performance, USSR
CIA estimates of, 136–38
effect of labor force participation on, 130–34
problems with CIA estimates of, 138
robustness of CIA estimates for, 143–46

Economic poverty, 28
Economic well-being
indicators in 1970s and 1980s of U.S., 27–28
infant mortality rate as indicator of, 29–33
suicide rates as indicator of, 33–34
Education
as factor in level of poverty, 59–60
as factor in prenatal health care, 60–61
less-developed country spending on, 231–232
role in South Africa for, 165–66
in Soviet Muslim republics, 132–33
See also Literacy; Illiteracy rates
Elderly population, Japan, 253–54
Elliot, David W.P., 277 n30
Environmental degradation, 249–50
Ethiopia, 89
Ethnic groups
in South Africa, 6–7, 168
in Soviet Union, 122
See also European ethnic population, USSR; Muslim population; Nationalities, USSR
European ethnic population, USSR, 122, 124–25, 257–60
Exports
developing country revenues from, 186, 188
revenues for sub-Saharan Africa from, 190

Faigenblum, J., 284 n23
Family planning, China, 251–52
Famine, 89
FAO. *See* Food and Agricultural Organization (FAO)
Fears, J. Rufus, 282 n18
Federal Republic of Germany (FRG)
death rates in, 80, 104, 108
economic growth in, 148
per capita output in, 145
spending, 11
Fertility
as force for population change, 250–51
Japanese subreplacement, 252–53
Fertility rates
comparison of Israeli and Palestinian, 255–56
projected levels of, 268
of Soviet Russian and Muslim population, 257–58
UN estimates for less-developed countries, 243
U.S. Census Bureau projections of USSR, 257–58
in USSR, 122–23
See also Family planning
Feshbach, Murray, 276 nn2, 3; 277 nn1, 2; 280 nn3, 15
Financial resources flow. *See* Borrowing, international; Capital flows; Debt, external; Official development assistance (ODA)
Fiscal transfers, less-developed countries, 236
Food and Agricultural Organization (FAO)
agricultural and nutritional data of, 170–71
estimates of undernutrition by, 172
Food consumption. *See* Consumption
Food economy, world
factors in creating crisis in, 193–94
factors influencing stabilization of, 185–89
instability in markets of, 194
performance of markets in, 185–86
reliable estimates of production in, 181–85
trends for, 189
See also Grain markets, international
Food production and supply
effect of misinformation about, 176
estimates in developing countries of, 174–76
variations and instability in data for, 180–84

Foreign aid. *See* Official development assistance (ODA)
France, 239–40
FRG. *See* Federal Republic of Germany (FRG)
Fuller, Elizabeth, 276 n10

Gadson, Kenneth E., 274 n5
German Democratic Republic (GDR)
 adult mortality rates in, 80, 99–102
 cigarette consumption in, 110–11
 death rates in, 104, 108
 distilled spirits consumption in, 111–14
 infant mortality rates in, 94–96
 life expectancy in, 78, 93–99
 per capita output in, 145–46
German Economic Research Institute (DIW), 146
GDR. *See* German Democratic Republic (GDR)
Glasnost, 92
Glewwe, Paul, 287 n32
Goering, T. J., 283 n15
Golladay, F., 284 n21
Government
 expanded size and power, 10–13
 factors in growth of, 14
 increased dependence on benefits of, 19
 post–World War II spending, 11
 use and misuse of statistics, 1–3, 15–16
 See also Public policy; Spending, government
Government intervention
 to change developing country private consumption, 201–3
 effect on international food markets of, 187–88, 194–95
 effect on world food markets of, 194–96
 increase in developing countries of, 237
 in sub-Saharan Africa, 190, 191, 192, 248
Grabauskas, V., 110, 278 n19
Grain markets, international
 futures markets in, 187–88
 prices in and performance of, 189, 192–95
Grain production, world, 171
Grand Apartheid, 169, 255
Great Leap Forward, 82–83, 89
Green Revolution, 180

Hadja, Lubomyr, 289 n39
Hagemann, Robert P., 289 n26
Hayek, Friedrich von, 22
Hazell, P. B. R., 284 nn25, 26
Health
 death rate as measure of, 157
 as deprivation indicator, 28–29
 deterioration in Communist countries of, 76–78, 92, 260–61
 less-developed country spending on, 231–33
Health and Nutrition Examination Survey (HANES), 172
Health care
 Communist government claims for improvement, 76
 perceived lack in United States of adequate, 44
 perception of spending by socialist governments, 82
 policy in Communist East-bloc countries for, 114–18
 prenatal and infant, 60–62
 U.S. consumer expenditures for, 69–73
 U.S. expenditures for high infant mortality group, 66–67
 See also Children; Disease; Life expectancy; Mortality rates, infant
Health status
 in Communist East-bloc countries, 99–101, 118–19
 differentiation in U.S. households for, 60–62
 related to alcohol consumption, 80–81, 111–14
 in USSR, 11–12, 126-30
Health systems
 in East-bloc countries, 81

in USSR, 76–78, 87
See also Life expectancy; Life expectancy at birth; Malnutrition; Medical system; Mortality rates, USSR
Heller, R. F., 278 n15
Heston, Alan, 283 n16
Hill, Kenneth, 277 n25
Homelands, South Africa
exclusion from census of, 152–53
independence for, 256
material conditions of population in, 166
Honey, P. J., 277 n31
Hong Kong, 51
Horiuchi, S., 101, 278 n13
Huber, Michel, 287 n3
Human capital, 247–48
Hungary
adult mortality rates in, 100–1
cigarette consumption in, 110–11
distilled spirits consumption in, 110–11
health care in, 114–18
infant mortality rates in, 94–96
life expectancy in, 78, 93–95, 96–98
mortality rates in, 79
per capita output in, 145
Hunger, 35

ICP. *See* International Comparison Project (ICP)
IFPRI. *See* International Food Policy Research Institute (IFPRI)
Illegitimacy
correlated with low birthweight, 56–59
as factor in infant mortality, 32–33, 34, 69
incidence in United States of, 61–62
rise in United States of, 42
in South Africa, 160
U.S., compared with rest of world, 62–65
See also Aid to Families with Dependent Children (AFDC)
Illiteracy rates
in Communist-governed countries, 89–91
for South Africans by racial designation, 165–66
IMF. *See* International Monetary Fund (IMF)
Income, personal, 65
Income-calorie relationship, 175
Income distribution estimates, 175
Indian Health Service (United States), 30
Industrialized countries
projected fertility rates in, 268
projected population growth in, 265–67
trend toward diminished status for, 269
Infant mortality. *See* Mortality rates, infant
Information networks, 188
International Comparison Project (ICP), 144–45
International Food Policy Research Institute (IFPRI), 173–74
International Monetary Fund (IMF)
data related to East-bloc countries, 11–12
estimates of government spending as ratio of GDP, 228–31
estimates for gross domestic investment in less-developed countries, 204
estimates of spending for health and education, 231
Investment
in less-developed countries, 199–201, 203, 204–5, 208–9
to middle-income countries, 207–9
in sub-Saharan Africa, 190–91
See also Capital flows
Irwig, L. M., 163, 282 nn13, 15
Israel, 254–55

Jablensky, A., 119, 280 n43
Japan
birth and fertility rates in, 251–54
elderly population in, 253
savings rate of, 253
Jevons, W. Stanley, 249
Johnson, D. Gale, 284 n24

Jolly, R., 287 n27

Kahn, Herman, 288 n17
Kelson, M. C., 278 n15
Kingkade, W. Ward, 257; 276 nn 6, 7; 282 nn5, 6, 11; 290 n52
Kinsella, Kevin, 290 n54
Kirner, W., 281 n12
Kleinman, Joel C., 275 n26
Klinger, A., 94–96, 277 n7
Komaros, Boris, 288 n18
Korea, North, 74
Kotas, Jiri V., 276 n16
Kravis, Irving B., 283 n16
Krueger, Anne O., 285 n2
Kuomintang, Taiwan, 7
Kuznets, Simon, 240, 287 nn1, 6; 290 nn50, 51

Labor force, USSR, 125, 130–34
Lamm, G., 107–8, 278 n17
Lampe, John, 288 n18
Lange, Oskar, 22
Lardy, Nicholas R., 288 n23
Lebanon, 254–55
Lelbach, Werner, 276 n12
Lenin, V.I., 4
Less-developed countries
- data development and collection in, 173
- direct, private investment in, 207
- food grain imports of, 185–86
- mortality rates as indicators of living conditions, 176-80
- national output and consumption of, 199–203
- projected fertility rates in, 268
- projected population growth in, 265–68
- reliability of statistics of, 199–205
- UN population growth estimates for, 243
- *See also* Children; Debt, external; Life expectancy; Mortality rates, infant

LIBOR. *See* London Interbank Offered Rates (LIBOR), rate loans
Life expectancy
- in Communist East-bloc countries, 78–79, 94–101, 261
- in Cuba, 85
- in developing countries, 178
- highest levels of U.S., 34
- improvements in U.S., 29
- increase in Sri Lanka of, 179
- infant mortality as determinant of in United States, 29–33
- in Japan, 259
- in South Africa, 157
- in sub-Saharan Africa, 178
- U.S., compared with other countries, 48–49
- in USSR, 77–78, 126–27, 129–30, 260
- *See also* Life expectancy at birth

Life expectancy at age one, 96
Life expectancy at birth
- in Communist East-bloc countries, 118
- as indicator of total national mortality, 93–94, 157
- in less-developed regions, 198
- for South African Whites, Asians, and Coloureds, 157–62
- U.S., compared with rest of world, 48–49
- in USSR, 77–78

Linked Birth and Infant Death Files (NCHS), 60
Lipton, Michael, 283 n8, 284 n18, 287 n30
Literacy in Communist countries, 89–91
Litmanowcz, M., 279 n38
Living conditions
- East-bloc health status as measure of, 23
- mortality rates as indicator of, 176–80
- perception of deterioration in low-income countries, 220–27

Loans, international. *See* Borrowing, international; Concessional loans; LIBOR (London Interbank Offered Rates), rate loans
London Interbank Offerred Rates (LIBOR), rate loans, 218–20
- *See also* Concessional loans

Lubin, Nancy, 280 nn8, 16; 289 n42

McCarthy, B. J., 274 n9
McGranahan, D., 284 n17
Maddison, Angus, 246; 249; 250 nn2, 5, 6; 275 nn20, 21; 288 n21
Malnutrition
 FAO assessment of developing country, 173–74
 FAO estimates of world, 171–72
 IFPRI measurement in developing countries of, 174–76
 incidence in developing countries, 174
 RDAs as basis for assessing, 37–38
 in USSR, 87
 See also Caloric intake and requirements; Undernutrition
Malthusian concept of population, 243
Mari Bhat, P. N., 284 n30
Material poverty, 28
Medical system, USSR, 128
Melia, R. J. W., 279 n27
Menarche age, 38–39
Mexico, 7
Migration
 as force for population change, 261–64
 forces precipitating, 262–63
 from South African urban areas, 154–57, 166
Military spending, USSR, 147
Miller, George, 274 nn3, 4
Mises, Ludwig von, 22
Miskiewicz, Sophia, 117, 279 nn24, 40
Mitchell, R. Judson, 280 n1
Mixed Marriage Act, South Africa, 256
Monetary policies, world, 193
Monthly Vital Statistics Report (MVSR), NCHS, 61–62
Morse, Barnett, 288 n17
Mortality
 as force for population change, 259–61
 life expectancy at birth as indicator of, 93–94
 neonatal, 96
 perinatal, 54–56
 WHO definition of infant, 50
 See also Life expectancy at birth
Mortality rates
 age-specific East-bloc, 99–101
 changing levels in USSR and Soviet bloc, 94–95, 260–61
 in Communist East-bloc countries, 77–81, 102–7, 119
 in developing countries, 176–80
 as indicators of living conditions, 176–80
 levels in USSR of, 261
 of U.S. compared with the rest of world, 48
Mortality rates, children
 among South African Whites, Asians, and Coloureds, 163
 among urban Blacks, 163
 in developing countries, 178
Mortality rates, infant
 in China, 82
 in Ciskei, South Africa, 162
 comparison of United States and other industrialized countries, 32
 correlation with poverty rates for children, 30–31
 correlation with U.S. poverty rate, 66
 in Cuba, 85
 data in less-developed countries for, 224
 definition of, 44–49
 in developing countries, 178, 198–99
 effect of parental behavior on U.S., 65–69
 estimates of UN Population Division, 221–24
 factors contributing to high U.S., 43–44
 factors influencing, 31–32
 in less-developed regions, 198
 levels in USSR of, 87, 127
 official estimates of, 221–24
 as social indicator, 29–33, 43
 for South African Blacks, 161–62
 for South African Whites, Asians, and Coloureds, 159–62

UNICEF interpretation of, 221
in United States, 32, 44–48, 69
U.S. compared with other countries, 43–49, 160
U.S. policy based on, 19–21
in USSR, 126–27, 261
See also Birthweight; Illegitimacy; Life expectancy at birth; Poverty rates, children
Mortality rates, USSR
causes of rising, 23, 127–28
changing levels of, 76–77, 126–30
consequences of rising, 128–30
levels of infant, 126–27, 261
WHO data for, 143
See also Life expectancy, USSR; Mortality rates, infant
Mugabe, Robert, 213
Mugford, Miranda, 274 nn12, 13, 14
Muslim population
fertility rates in USSR of, 122–25, 257–60
in Soviet labor force, 131–34
in USSR, 122–23
MVSR. *See Monthly Vital Statistics Report (MVSR)*, NCHS

Nagle, R. F., 282 n13
National Center for Health Statistics (NCHS), 21, 29, 40
National Food Consumption Survey (1977–1978), 36, 37
Nationalities, USSR, 257–58
National Pact, Lebanon, 254
National security and immigration, 263–64
NCHS. *See* National Center for Health Statistics (NCHS)
Netherlands, The, 46, 51
New Zealand, 46, 48
Nhan Dan, 277 n30
Nicoletti, Guiseppe, 289 n26
Nutrition
as deprivation indicator, 28
effect of improvement in, 179
FAO estimates of, 171
measures for assessment of adequacy of, 35–40
See also Caloric intake and requirements; Malnutrition; Obesity; Undernutrition
Nutritional deprivation, U.S., 38–40
Nutritional status
in Communist-governed countries, 87–89
decline in vulnerability of, 189
role of government in maintaining, 195–96
for South African population, 163–65
in sub-Saharan Africa, 189

Obesity, 40, 176
ODA. *See* Official development assistance
OECD. *See* Organization for Economic Cooperation and Development (OECD)
Official development assistance (ODA), 25, 197–99
Organization for Economic Cooperation and Development (OECD)
estimates of disbursements and transfers to low-income countries, 205–7, 215–16
estimates of disbursements to newly industrializing countries, 207–8
estimates of infant mortality, 44–48
Organski, A. F. K., 264–65, 290 n53
Ostfeld, Adrian, 279 n42

Passell, Jeffrey S., 274 n7, 283 n7
Paulino, A., 283 n5
Pauperism, 42
PCE. *See* Personal consumption expenditure (PCE)
Pear, Robert, 281 n3
People's Republic of China (PRC). *See* China
Personal consumption expenditure (PCE), 37, 164
Peru, 235
Petersen, William, 287 n9
Poland
adult mortality rates in, 99–101
cigarette consumption in, 110–11

distilled spirits consumption in, 111–14
government spending (1981–1985), 11–12
health care in, 115–17
infant mortality rates in, 94–96
life expectancy in, 92–94, 96–99
per capita output in, 145
Population
projections for and redistribution of global, 264–69
projections for North America, Latin America, and Africa, 241–42
Population, South Africa
adjustment of composition of, 152–56
effect of distribution of, 152–56
See also Homelands, South Africa
Population, USSR
effect of rising mortality rates on, 128–30
projected size of, 121, 125
shifts in composition of, 120–25, 257–59
Population Classification Act, South Africa, 256
Population growth
estimates of U.S., 241–42
as form of social change, 249
misconceptions and truths about low-income countries', 243–51
policy in China for, 82–84
presumed effect in less-developed countries of, 243–44
projections for, 241–42, 265–66
in sub-Saharan Africa, 178–79, 247
UN estimates of world, 242
Poverty
under communism, 91
Communist government claims in combating, 75
existence and measurement of U.S., 42, 51–54
prevalence in United States of, 44
of South African Bantustans, 142
South African government policy for, 124-26, 140, 142
spending of U.S. household in, 36–38, 52
U.S. children in, 44, 51–52
of whites in South Africa, 124–25, 140
See also Antipoverty policy, South Africa
Poverty index, 41, 52
Poverty rate, U.S., 16–19, 27, 29, 52, 59–60
consumption-based, 18–19
income-based, 17
See also Economic poverty; Material poverty
Poverty rates, children
correlation with infant mortality, 30–31
levels for black, 31
in United States, 51–52
Powell-Greiner, Eve, 275 n23
PRC (People's Republic of China). *See* China
Preston, Samuel, 110, 278 n22, 284 n30
Prewitt, Kenneth, 21
Prices, international, 186–87, 192–93
Productivity, USSR (CIA estimates), 140–41
Property rights, South Africa, 256
Purchasing power, nutritional, 189
Purchasing power parity, 144

Rapawy, Stephen, 277 n1
Rationalism, 10
RDAs. *See* Recommended dietary allowances (RDAs)
Reagan administration, 12–13
Recommended dietary allowances (RDAs), 37–38
Republic of South Africa (RSA). *See* South Africa
Reutlinger, S., 284 n19
Robinson, J. Gregory, 274 n7, 283 n7
Romania
adult mortality rates in, 99–101
cigarette consumption in, 110–11
distilled spirits consumption in, 111–12
government spending (1981–1985), 12
health care in, 114–18

infant mortality rates in, 95–96
life expectancy in, 93
Roof, Michael, 289 n36
Roosevelt, Theodore, 249
Rowen, Henry S., 281 n2
RSA. *See* South Africa
Russian ethnic group. *See* European ethnic population, USSR
Ruttan, Vernon W., 285 n2

Schiff, Gary, 289 n32
Schneider, R. E., 284 n23
Schultz, Theodore W., 247, 287 n29
Schumpeter, Joseph, 273 n1
Schwartz, Marvin, 286 n9
Self-determination, 4–5
Shiffman, M., 284 n23
Silver, Brian, 256, 277 n29, 289 nn39, 40
Simon, Julian, 288 n17
Slesnick, Daniel T., 17
Smallwood, David M., 274 n5
Smil, Vaclav, 288 n18
Smoking
in Communist East-bloc countries, 110–11
as factor affecting birthweight, 57
Social welfare programs, 34
See also Dependency indicator
South Africa
demographic policy of, 256–57
infant mortality in, 159–62
life expectancy by race in, 157–59
nutritional status of people in, 163–65
policy for Afrikans population, 150–52, 168–69
relaxed racial laws in, 256–57
See also Population, South Africa
Spending, government
East-bloc countries (1981–1985), 11–12
evaluation of data for, 229–36
IMF estimates of adjustments in, 228–30
as measure of government expansion, 11–13
on poor people in U.S., 16
for public health in Communist-bloc countries, 81–82
for South African education, 165–66
for South African social welfare, 166–68
See also Adjustment policy; Borrowing, international; Military spending, USSR; State-owned enterprises; Subsidies
Spending, household
in United States, 35–38
of U.S. poor, 52
Spengler, Joseph, 287 nn4, 5
Stalin, Josef, 76
State-owned enterprises
in developing countries, 235
in sub-Saharan Africa, 247–48
Statistics. *See* Data
Stewart, F., 287 n27
Sub-Saharan Africa
economic problems of, 189–91, 248
fallacy of population growth argument in, 248
Subsidies, less-developed countries, 235
Suicide rates, 33–34
Summers, Robert, 283 n16
Swan, A. V., 279 n27
Sweden, 13

Taeuber, Cynthia, 290 n54
Taiwan, 7
TBVC (Transkei, Bophuthatswana, Venda, Ciskei) population, 152–53, 156
Terreblanche, S. J., 151, 282 n4
Tilly, Charles, 241, 287 n8
Tongas, Gerard, 89–90, 277 n31
Torrey, Barbara B., 290 n54
Transkei, 152
Treml, Vladimir, 276 nn11, 12
Truman, Harry S, 197, 205
Truss, T. J., 279 n42
Tseng, S. S., 283 n5

Uemura, K., 93, 277 n5
Undernutrition, 171–80

UN Children's Fund (UNICEF)
assessment of condition of world's children, 212–14, 221, 225–27, 236
assessment of health and education programs, 230
assessment of school enrollment, 226–27
criticism of adjustment policies, 228
evaluation of Peru's subsidy structure, 235–36
interpretation of third world debt crisis, 25
UNESCO. *See* UN Educational, Scientific, and Cultural Organization (UNESCO) data
UN Educational, Scientific, and Cultural Organization (UNESCO) data, 224–26
UNFPA. *See* UN Fund for Population Growth (UNFPA)
UN Fund for Population Growth (UNFPA), 244
UNICEF. *See* UN Children's Fund (UNICEF)
UN Population Division
estimates of infant mortality rates, 44–48, 221–24
estimates of life expectancy at birth, 93–94
estimates of school enrollment, 227
life expectancy estimates for USSR, 126
population growth estimates of, 243
population projections of, 243
UN Statistical Office
estimates of output and structure, 232–33
International Comparison Project (ICP), 144–45
United Kingdom, 13
Urbanization pattern, South Africa, 154–56
U.S. Agency for International Development (USAID), 244
U.S. Bureau of the Census
estimates of USSR life expectancy, 127
projection of fertility rates in USSR, 124, 257
U.S. Presidential Commission on World Hunger (1980), 173
USAID. *See* U.S. Agency for International Development (USAID)
Usher, Dan, 28

van Wolferen, Karel, 290 n56
Venda, 152
Vietnam, 74, 89
Vishnevskiy, Anatoliy, 277 n3, 281 n7

Warsaw Pact nations. *See* Communist East-bloc countries
Wealth
perception of transfers from poor to rich countries, 214–20
transfers from industrial to less-developed countries, 205–8
Weather, sub-Saharan Africa, 190, 191
West Germany. *See* Federal Republic of Germany (FRG)
WHO. *See* World Health Organization (WHO)
Wiemer, Reinhard, 289 n35
Wolf, Charles, Jr., 281 n2
World Bank
estimates of developing country investment, 199–201, 204–5
estimates of net financial transfers, 216–17
reliability and accuracy of economic data, 199–204
World Health Organization (WHO)
MONICA project, 108
USSR mortality data, 143
World War I, 4–5
World War II, 4–5
Wyndham, C. H., 161, 162, 163, 282 n15

Yeh, K. C., 284 n24
Yugoslavia
death rates in, 105, 109
infant mortality rates in, 96
life expectancy in, 93–94
per capita output in, 145

Zaichenko, Aleksandr, 281 n4

About the Author

Nicholas Eberstadt is a visiting scholar at the American Enterprise Institute and a visiting fellow at the Center for Population Studies, Harvard University. He has been a consultant for such agencies as the U.S. Agency for International Development, U.S. Bureau of the Census, U.S. Department of State, and World Bank.

The author has taught courses at Harvard University in population and natural resources, agricultural economics, social science and social policy, and problems of policy making in less-developed countries.

Of his many books and articles, the two latest are *US Foreign Aid Policy—A Critique* (Foreign Policy Association, 1990) and *The Population of North Korea* (University of California Press, 1992).

Credits

Chapter 1: *Public Interest,* Winter 1988. Reprinted with permission.

Chapter 2: *Public Interest,* Fall 1991. Reprinted with permission. Research for this chapter was supported by a generous grant from the Smith Richardson Foundation.

Chapter 3: *Journal of Economic Growth,* vol. 2, no. 2, 1987. Reprinted with permission.

Chapter 4: *Communist Economies,* vol. 2, no. 3, 1990. Reprinted with permission.

Chapter 7: *Optima* magazine, March 1988. Reprinted with permission.

Chapter 8: Previously unpublished. Research for this chapter was supported by a generous grant from the Lynde and Harry Bradley Foundation.

Chapter 9: *Journal of Economic Growth,* Summer 1989. Reprinted with permission.

Chapter 10: Testimony delivered before Senate Foreign Relations Committee, April 5, 1990. Published in an abridged form by the *Journal of Economic Growth,* December 1990.

Chapter 11: Abridged form appeared in *Foreign Affairs,* Summer 1991.

The American Enterprise Institute for Public Policy Research

Founded in 1943, AEI is a nonpartisan, nonprofit, research and educational organization based in Washington, D.C. The Institute sponsors research, conducts seminars and conferences, and publishes books and periodicals.

AEI's research is carried out under three major programs: Economic Policy Studies; Foreign Policy and Defense Studies; and Social and Political Studies. The resident scholars and fellows listed in these pages are part of a network that also includes ninety adjunct scholars at leading universities throughout the United States and in several foreign countries.

The views expressed in AEI publications are those of the authors and do not necessarily reflect the views of the staff, advisory panels, officers, or trustees.

A NOTE ON THE BOOK

This book was edited by Ann Petty of the publications staff of the American Enterprise Institute. The index was prepared by Shirley Kessel. The text was set in Palatino, a typeface designed by the twentieth-century Swiss designer Hermann Zapf. Publication Technology Corporation, of Fairfax, Virginia, set the type, and Edwards Brothers Incorporated, of Lillington, North Carolina, printed and bound the book, using permanent acid-free paper.

The AEI PRESS is the publisher for the American Enterprise Institute for Public Policy Research, 1150 17th Street, N.W., Washington, D.C. 20036; *Christopher DeMuth*, publisher; *Dana Lane*, director; *Ann Petty*, editor; *Leigh Tripoli*, editor; *Cheryl Weissman*, editor; *Lisa Roman*, editorial assistant (rights and permissions).